Communications in Computer and Information Science 1464

More information about this series at http://www.springer.com/series/7899

Jinsong Su · Rico Sennrich (Eds.)

Machine Translation

17th China Conference, CCMT 2021
Xining, China, October 8–10, 2021
Revised Selected Papers

 Springer

Editors
Jinsong Su
Xiamen University
Xiamen, China

Rico Sennrich
The University of Edinburgh
Edinburgh, UK

ISSN 1865-0929   ISSN 1865-0937 (electronic)
Communications in Computer and Information Science
ISBN 978-981-16-7511-9   ISBN 978-981-16-7512-6 (eBook)
https://doi.org/10.1007/978-981-16-7512-6

This Springer imprint is published by the registered company Springer Nature Singapore Pte Ltd.
The registered company address is: 152 Beach Road, #21-01/04 Gateway East, Singapore 189721, Singapore

Preface

The China Conference on Machine Translation (CCMT) is a national annual academic conference held by the Machine Translation Committee of the Chinese Information Processing Society of China (CIPSC), which brings together researchers and practitioners in the area of machine translation, providing a forum for those in academia and industry to exchange and promote the latest developments in methodologies, resources, projects, and products, with a special emphasis on the languages in China. Since the first session of CCMT in 2005, 16 sessions have been successfully organized (the previous 14 sessions were called CWMT), and a total of 10 machine translation evaluations (2007, 2008, 2009, 2011, 2013, 2015, 2017, 2018, 2019, 2020) have been organized, as well as one open source system module development (2006) and two strategic seminars (2010, 2012). These activities have made a substantial impact on advancing the research and development of machine translation in China. The conference has been a highly productive forum for the progress of this area and is considered a leading and important academic event in the natural language processing field in China.

This year, the 17th CCMT took place in Haidong, Qinghai. This conference continued the tradition of being the most important academic event dedicated to advancing machine translation research in China. It hosted the 11th Machine Translation Evaluation Campaign, featured two keynote speeches delivered by Minlie Huang (Tsinghua University) and Lei Li (University of California), and included two tutorials delivered by Zhongjun He (Baidu) and Liangyou Li (Huawei). The conference also organized five panel discussions, bringing attention to pre-training and machine translation, end-to-end speech translation, the industry of machine translation, the frontier of machine translation, and the forum for PhD students. A total of 85 submissions (including 25 English papers and 60 Chinese papers) were received for the conference. All papers were carefully reviewed in a double-blind manner and each paper was evaluated by at least three members of an international Program Committee. From the submissions, 10 English papers were accepted. These papers address all aspects of machine translation, including improvement of translation models and systems, translation quality estimation, document-level machine translation, low-resource machine translation, etc. We would like to express our thanks to every person and institution involved in the organization of this conference, especially the members of the Program Committee, the machine translation evaluation campaign, the invited speakers, the local organization team, our generous sponsors, and the organizations that supported and promoted the event. Last but not least, we greatly appreciate Springer for publishing the proceedings.

August 2021

Jinsong Su
Rico Sennrich

Organization

General Chair

Nyima Tashi   Tibet University, China

Program Committee Co-chairs

Jinsong Su   Xiamen University, China
Rico Sennrich   The University of Edinburgh, UK

Evaluation Chairs

Muyun Yang   Harbin Institute of Technology, China
Yating Yang   Xinjiang Technical Institute of Physics and Chemistry,
            Chinese Academy of Sciences, China

Organizing Co-chairs

Rangjia Cai   Qinghai Normal University, China
Jiajun Zhang   Institute of Automation, Chinese Academy of Sciences,
            China

Tutorial Co-chairs

Yang Feng   Institute of Computing Technology, Chinese Academy
            of Sciences, China
Rui Wang   Shanghai Jiao Tong University, China

Student Forum Co-chairs

Chong Feng   Beijing Institute of Technology, China
Changliang Li   Kingsoft, China

Front-Trends Forum Co-chairs

Shujian Huang   Nanjing University, China
Tong Xiao   Northeastern University, China

Workshop Co-chair

Hui Huang   University of Macau, China

Publication Co-chair

Yanqing He   Institute of Scientific and Technical Information
                              of China, China

Sponsorship Co-chairs

Guoping Huang   Tencent, China
Cunli Mao   Kunming University of Science and Technology, China

Publicity Co-chairs

Junhui Li   Soochow University, China
Hao Yang   Huawei, China

Program Committee

Hailong Cao   Harbin Institute of Technology, China
Kehai Chen   NICT, Japan
Yidong Chen   Xiamen University, China
Yufeng Chen   Beijing Jiaotong University, China
Yong Cheng   Google, USA
Cunli Mao   Kunming University of Science and Technology, China
Jinhua Du   Investments AI, AIG, UK
Xiangyu Duan   Soochow University, China
Yang Feng   Institute of Computing Technology, Chinese Academy
                              of Sciences, China
Xiaocheng Feng   Harbin Institute of Technology, China
Shengxiang Gao   Kunming University of Science and Technology, China
Zhengxian Gong   Soochow University, China
Zhongjun He   Baidu, China
Yanqing He   Institute of Scientific and Technical Information
                              of China, China
Guoping Huang   Tencent, China
Shujian Huang   Nanjing University, China
Degen Huang   Dalian University of Technology, China
Yves Lepage   Waseda University, Japan
Junhui Li   Soochow University, China
Maoxi Li   Jiangxi Normal University, China
Liangyou Li   Huawei, China
Yachao Li   Soochow University, China
Hui Li   Xiamen University, China
Xiang Li   Xiaomi, China
Yang Liu   Tsinghua University, China
Qun Liu   Huawei, China
Lemao Liu   Tencent, China

Fandong Meng	Tencent WeChat, China
Haitao Mi	Ant Financial, USA
Toshiaki Nakazawa	Kyoto University, Japan
Kai Song	Alibaba, China
Linfeng Song	Tencent, USA
Jinsong Su	Xiamen University, China
Xu Tan	Microsoft, China
Zhaopeng Tu	Tencent, China
Xing Wang	Tencent, China
Shaonan Wang	Institute of Automation, Chinese Academy of Sciences, China
Mingxuan Wang	ByteDance, China
Rui Wang	Shanghai Jiao Tong University, China
Longyue Wang	Tencent, China
Hui Huang	University of Macau, China
Shuangzhi Wu	Tencent, China
Changxing Wu	East China Jiaotong University, China
Xiaofeng Wu	Apple, UK
Tong Xiao	Northeastern University, China
Jun Xie	Alibaba, China
Hongfei Xu	Saarland University, Germany
Jinan Xu	Beijing Jiaotong University, China
Muyun Yang	Harbin Institute of Technology, China
Baosong Yang	Alibaba, China
Yating Yang	Xinjiang Institute of Physics and Chemistry, Chinese Academy of Sciences, China
Heng Yu	Alibaba, China
Jiajun Zhang	Institute of Automation, Chinese Academy of Sciences, China
Dakun Zhang	Systran, France
Wen Zhang	Tencent, China
Biao Zhang	University of Edinburgh, UK
Haibo Zhang	Alibaba, China
Hao Zhou	ByteDance, China
Muhua Zhu	Alibaba, China

Organizer

Chinese Information Processing Society of China, China

Co-organizer

Qinghai Normal University, China

Sponsors

Diamond Sponsors

Global Tone Communication Technology Co., Ltd.

Volctrans

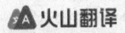

Platinum Sponsors

NiuTrans Research

Youdao

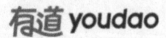

NEWTRANX Technology

Transn

Tencent TranSmart

Gold Sponsors

Xiaomi Corporation

Baidu

Cloud Translation

Silver Sponsor

Jeemaa

Contents

A Document-Level Machine Translation Quality Estimation Model Based on Centering Theory

Yidong Chen[1,2(✉)], Enjun Zhong[1,2], Yiqi Tong[1,2,3], Yanru Qiu[1,2], and Xiaodong Shi[1,2]

[1] Department of Artificial Intelligence, School of Informatics, Xiamen University, Xiamen, China
[2] Key Laboratory of Digital Protection and Intelligent Processing of Intangible Cultural Heritage of Fujian and Taiwan, Ministry of Culture and Tourism, Xiamen, China
[3] Institute of Artificial Intelligence, Beihang University, Beijing, China
{ydchen,mandel}@xmu.edu.cn, {ejzhong,yqtong,yrqiu}@stu.xmu.edu.cn

Abstract. Machine translation Quality Estimation (QE) aims to estimate the quality of machine translations without relying on golden references. Current QE researches mainly focus on sentence-level QE models, which could not capture discourse-related translation errors. To tackle this problem, this paper presents a novel document-level QE model based on Centering Theory (CT), which is a linguistics theory for assessing discourse coherence. Furthermore, we construct and release an open-source Chinese-English corpus at https://github.com/ydc/cpqe for document-level machine translation QE, which could be used to support further studies. Finally, experimental results show that the proposed model significantly outperformed the baseline model.

Keywords: Machine translation · Document-level quality estimation · Centering theory

1 Introduction

Machine translation quality estimation (QE) is a task that aims at automatically estimating the quality of machine translations. Unlike the standard evaluation metrics such as BLEU [15], NIST [4] and METEOR [1], QE models estimate translations without relying on golden references. In the past decade, researches on QE have attracted more and more attentions [7], since QE can be utilized to ensure the diversity and robustness of the NMT systems [25].

Currently, mainstream QE-related researches [2,13,26] mainly focus on sentence-level QE models, which normally ignore the document-level information. While, previous studies [21,23] have shown that document-level information is important for estimating the translation qualities. As shown in Fig. 1, the word

J. Su and R. Sennrich (Eds.): CCMT 2021, CCIS 1464, pp. 1–15, 2021.
https://doi.org/10.1007/978-981-16-7512-6_1

Context source: 五月份，俄罗斯农业监督局宣布，本国农业年度对华粮食出口首次超过100万吨，创新记录。
Context translation: In May, the Russia Agricultural Supervision Bureau announced that for the first time in the agricultural year, Russia's grain exports to China exceeded 1 million tons, setting a new record.

Current source: 该局预测中国可能进入前十大俄罗斯粮食进口国之列。
Current translation: The bureau predicts that China may be among the top ten Russian food importers.

Fig. 1. An example of a translation that is correct in sentence-level but incorrectly in document-level. We use THUMT [20] and 2M Chinese-English parallel data to training the NMT model.

"predicts" in current translation should be "predicted" according to the context, but is wrongly translated into present tense. Obviously, a QE model that does not consider the document-level information could not tell the above-mentioned error.

To alleviate this problem, we propose a document-level QE model called CpQE by introducing Centering Theory (CT) [24] to formulate the sentence relations. Concretely, our CpQE model uses the Preferred Center (Cp), whose meaning could be found in Subsect. 3.1, to represent the context features. Moreover, we adapt a BERT-based [3] sequence labeling model to extract the Cps. In addition, a semi-supervised pseudo-label learning method is adopted to alleviate the low resource problem of Cp extraction.

2   Related Work

Traditional QE works [6, 17] used feature engineering to extract features, e.g. QuEst++ [19] design word-, sentence- and document-level features for multi-level QE. Recently, neural QE methods outperformed these hand-craft methods. [16] treated QE as a slot filling problem and proposed a language independent word-level QE system using Recurrent Neural Network (RNN). [14] proposed a stacked model by introducing multi-task learning, which achieved the best result for word-level and sentence-level QE at that time.

More recently, Predictor-Estimator framework [10] was reported superior performance and become a mainstream approach for neural QE. To combine Predictor and Estimator into the architecture, [13] proposed a unified neural network, which were trained jointly to minimize the mean absolute error over the QE training samples. Furthermore, [5] proposed a neural bilingual expert model, which replaced the RNN layers with a novel bidirectional transformer [22] for feature extraction. And [11] apply the pre-trained model, BERT [3], as feature extractor. However, these methods evaluate each translation independently, leading to an inconsistent problem for the evaluation of document-level machine translation.

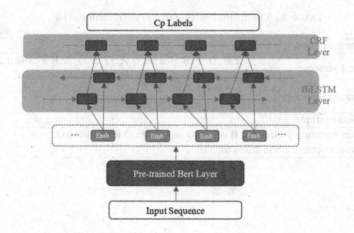

Fig. 2. The overview of Preferred Centering extraction model

3 Centering Theory and Extraction of the Preferred Centers

3.1 Centering Theory and Preferred Centers

Centering Theory (CT) [8,9,24] is a theoretical model about the local coherence of discourses. CT, which can be parameterized and calculated easily compared with other related theories, provides a quantitative standard for evaluating the context consistency of translations. Therefore, in this work, we apply CT to capture the discourse coherence information for document-level QE.

In CT, any entity in a sentence may relate to entities in the following sentences. So an entity is called Forward-looking Center (Cf). And an entity related to entities in the previous sentences is called Backward-looking Center (Cb). Preferred Center (Cp) is the entity that is the most likely one to be associated with a Cb. For example, given a current sentence "Xiao Hong likes to wear a red skirt" and the following sentence "She went shopping today and met Xiao Fang". The entities in the current sentence include "Xiao Hong" and "skirt", so we have Cf = ["Xiao Hong", "skirt"]; and the Cb in following sentence is "she", i.e. Cb = ["she"]. In Cf, the word "Xiao Hong" is the most closely related to the Cb, so "Xiao Hong" is defined as the preferred center. It should be noted that a sentence may contains more than one Cps.

3.2 The Preferred Centers Extraction Model

The conventional methods for extracting Cp are mainly rule-based. While, in this paper, we take this problem as a sequence labeling problem and construct a BERT-BiLSTM-CRF based model to settle it.

Figure 2 presents the overview of our extraction model. The input sentences are encoded by BERT first. Then, the output of BERT are fed to a BiLSTM layer, in which the operations of the LSTM are shown as follows:

Table 1. The format of preferred center annotation.

Chinese example	
current sentence	小_B 明_I 和_O 小_B 红_I 决_O 定_O 去_O 电_B 影_I 院_I ._O
	(Xiao Ming and Xiao Hong decided to go to the cinema)
following sentence	他们看了一场精彩的电影
	(They watched a wonderful movie.)
English example	
current sentence	Brennan_B drives_O an_O Alfa_O Romeo_O ._O
following sentence	She drives too fast.

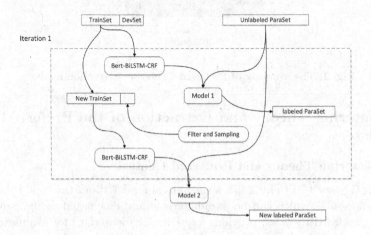

Fig. 3. The pipeline of our semi-supervised training method

$$i_t = \sigma(W_i[h_{t-1}, x_t] + b_i), \tag{1}$$

$$f_t = \sigma(W_f[h_{t-1}, x_t] + b_f), \tag{2}$$

$$c_t = f_t c_{t-1} + i_t \tanh(W_c[h_{t-1}, x_t] + b_c), \tag{3}$$

$$o_t = \sigma(W_o[h_{t-1}, x_t] + b_o), \tag{4}$$

$$h_t = o_t \tanh(c_t), \tag{5}$$

where x_t represents the output of BERT. i_t, f_t and c_t are the input gate, forget gate and cell vectors, respectively. o_t is the output gate and h_t is the hidden vector. t represents the t-th cell state of LSTM.

After that, the output of the forward and the backward LSTM are concatenated using (6), as follows:

$$h_t = [\overrightarrow{h_t}, \overleftarrow{h_t}] \tag{6}$$

Finally, the outputs of BiLSTM are provided to Conditional Random Field (CRF) [12] to decode the Cp labels.

3.3 The Semi-supervised Preferred Center Extraction Method

Since there are no public datasets for Cp extraction, we manually annotated a small-scale Cp extraction dataset. Concretely, the English corpus is annotated in word-level while the Chinese corpus is annotated in character-level. Table 1 shows the format of annotation. Considering that such a small annotated dataset is not enough for training a automatic annotation model, we proposed a semi-supervised method to do so. The training pipeline is shown in Fig. 3.

First, we divided the annotated dataset into training set and development set. Then we trained the BERT-BiLSTM-CRF model with these two sets to get Model 1. After that, we predict the unlabeled parallel corpus with Model 1 to get a labeled dataset. Next, we filtered the labeled data by rules to alleviate the effect of noise. Here are the rules we define:

- Remove the sentences whose ratio of the total length of preferred centers to the total length of sentence is more than 1/4.
- Calculate the maximum similarity between each preferred center and the words in the following sentence. If the similarity is less than 0.5 and such preferred center do not belong to any component of subject, direct object or indirect object, record this preferred center. If the number of such kind of preferred center is greater than or equal to 50% of the number of preferred centers extracted from the sentence, the sentence will be removed.

Roughly, Rule 1 limits the number of preferred centers to avoid selecting excessive entities as the preferred centers for higher recall, and Rule 2 remove the samples which contain ambiguous Cp. For measuring the similarity between words, we use a word2vec model[1] to encode the words into vectors and calculate their cosine similarity:

$$similarity(w_i, w_j) = \frac{emb_i * emb_j}{||emb_i|| * ||emb_j||} \tag{7}$$

where emb_i is the vectorized representation of w_i. If the out-of-vocabulary word can not be found in the following sentence, the similarity is set to be 0, otherwise 1.

After filtering the labeled dataset, the dataset will be randomly sampled to get three sampling datasets. These three datasets will be combined with the initial training set respectively for training three new models. Then we choose the highest recall model on development set as Model 2. Our goal is to obtain comprehensive preferred centers as far as possible so we choose the recall to select the optimal model. So far, we have completed one iteration. The next step is to repeat the previous steps.

4 The Quality Estimation Model

In this section, we present our CT-based document-level QE model. As shown in Fig. 4, we extract the features of preferred centers from two aspects by outer-extractor. First, we get the embeddings of preferred centers in both source and

---

[1] https://radimrehurek.com/gensim/models/word2vec.html.

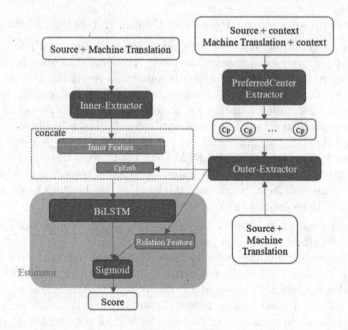

Fig. 4. The overview framework of our CpQE model

target side. Second, compute the consistency between current sentence and context in both source and target side. Finally, the two types of features and the inner sentence features extracted by inner-extractor are passed to the quality evaluator for scoring.

4.1   The Inner-Extractor

As shown in Fig. 5, the encoder of inner-extractor is a standard encoder of transformer [22] and the decoder is bidirectional. The forward self-attention network decodes the target words from left to right, while the backward self-attention network decodes the target words from right to left. The combination of the two self-attention can make the model focus on the whole sentence.

4.2   The Outer-Extractor

Outer-Extractor extract Cp features from two aspects: sentences relation features and embeddings of preferred centers. Sentences relation features can evaluate the coherence between source text and translations. Here we define four rules for designing features:

- The number of preferred centers of current sentence in source and target side and the difference between the numbers.
- The number of preferred centers of previous sentence in source and target side and the difference between the numbers.

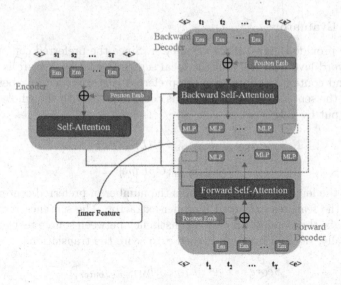

Fig. 5. The architecture of inner-extractor

- The similarity between preferred centers of previous sentence and current sentence in source and target side and the difference between the similarities.
- The similarity between preferred centers of previous sentence and preferred centers of current sentence in source and target side and the difference between the similarities.

Rule 1 and rule 2 focus on the number of preferred centers which can reflect the consistency between source text and translation at some extent. Rule 3 use a quantitative measurement to evaluate the consistency between previous sentence and current sentence. Rule 4 measure the change of entities which reflects the change of topic. If a sentence at the beginning of document, the preferred center of the previous is empty set. The preferred center of the last sentence in document is empty set too. The similarity between the sequence is computed as follow:

$$similarity(l_1, l_2) = \frac{l_{v1} + l_{v2}}{L_1 + L_2} cosine(emb_{w_{v1}}, emb_{w_{v2}})$$
$$+ \frac{2}{L_1 + L_2} \sum_{w in w_{o1}} f(w, l_2) \qquad (8)$$

$$f(w, l_2) = \begin{cases} 1, w\ in\ l_2, \\ -1, w\ not\ in l_2. \end{cases} \qquad (9)$$

where w_{v1} is the word in the sequence 1 which can be found in vocabulary while w_{o1} is the word in the sequence 1 which out of the vocabulary. l_{v1} is the length of w_{v1} and L_1 is the length of the sequence 1. w_{v1} and w_{v2} are calculated by Word2Vec model. According to the four rules, we design 12 features to represent sentence relation information. We provide the running process of outer-extractor on Appendix A.

4.3   The Evaluator

Finally, we provide the features to evaluator. Since the preferred center embedding is a word-level feature, and the local sentence relation feature is for both sentence and context, we integrate the preferred center embedding before BiLSTM. And the sentence relation feature is concatenated with the whole sentence feature output by BiLSTM:

$$\overrightarrow{h_{1:T+n}}, \overleftarrow{h_{1:T+n}} = BiLSTM(f) \tag{10}$$

$$f = [f_{inner}; CpEmb] \tag{11}$$

where T is the length of translation, n is the number of preferred centers. f_{inner} represents the features extracted by inner-extractor. The sentence relation feature can make the evaluator focus on consistency between source text and translation. Finally, sigmoid function is used σ to score the translations:

$$Score = \sigma(w^T[\overrightarrow{h_{1:T+n}}; \overleftarrow{h_{1:T+n}}; f_{outer}]) \tag{12}$$

where w is a trainable parameters, f_{outer} is the features extracted by outer-extractor. The optimization object is calculate as follows:

$$argmin||HTER - Score||_2^2 \tag{13}$$

$$HTER = \frac{N_{edit}}{N_{reference}} \tag{14}$$

where N_{edit} is the number of edits from translation to reference, $N_{reference}$ is the number of words in reference. Human-targeted Translation Edit Rate (HTER) [18] is the widest used metric of QE. Calculation of HTER need to find out the closest reference of the translation, then calculate the edit rate from translation to reference.

5   Experiments

5.1   Metrics

For preferred centers extraction, our goal is to maximize the total number of preferred centers that are correctly tagged by our method, so we use standard Accuracy and Recall score[2] to measure the performance of our BERT-based extraction model.

For quality estimation model, following with previous works such as [5, 14], we use Pearson correlation coefficient, which is calculated as follows.

$$\rho_{X,Y} = \frac{\sum_{i=1}^{n}(x_i - \mu_X)(y_i - \mu_Y)}{\sqrt{\sum_{i=1}^{n}(x_i - \mu_X)^2 \sum_{i=1}^{n}(y_i - \mu_Y)^2}} \tag{15}$$

Where n is the number of samples, μ_X and μ_Y denote means of the samples. A larger coefficient represents that X and Y are more correlated.

---

[2] https://github.com/chakki-works/seqeval

Table 2. Preferred center extraction performance

Chinese model	Recall	Accuracy	Training set
Rule base model	38.26%	34.53%	–
Model 1	51.74%	47.18%	1000 labeled data
Model 2	57.01%	53.83%	1000 labeled data + 1000 pseudo labeled data
Model 3	60.70%	59.44%	1000 labeled data + 1500 pseudo labeled data
English model	Recall	Accuracy	Training set
Rule based model	40.43%	39.17%	–
Model 1	53.09%	49.32%	1000 labeled data
Model 2	56.84%	56.28%	1000 labeled data + 1000 pseudo labeled data
Model 3	63.61%	61.08%	1000 labeled data + 1500 pseudo labeled data

Table 3. Pearson correlation coefficient of models. CpQE+CpRuled represents the preferred centers are extracted by rule. CpQE+CpSeq represents the preferred centers are extracted by our sequence labeling model.

Model	Sentence testset	Document testset
Baseline	0.6392	0.5536
CpQE+CpRuled	0.6218	0.5911(+0.0375)
CpQE+CpSeq	0.6326	0.6035(+0.0499)

5.2   Dataset Description

Since the lack of document-level QE corpus, we manually annotated an open source Chinese-English document-level dataset[3]. Concretely, our document-level QE corpus is built from the test set of WMT2019 MT automatic evaluation task. We select 996 Chinese source sentences from the corpus, including 112 articles with a text length less than 14 sentences, and the corresponding 1992 sentences of English translations. The 1992 translations are calculated the HTER value to construct our corpus.

For the preferred center extraction experiment, we use our annotated preferred center extraction dataset including 1,432 Chinese sentences and 1,432 English sentences. The Chinese-English parallel corpus comes from FBIS corpus including 10,355 documents and 228,611 sentence pairs are used to generate pseudo labeled data.

For the quality estimation experiment, we use CCMT19 Chinese-English sentence-level translation quality estimation dataset with 11,213 sentences and

---

[3] Available at https://github.com/ydc/cpqe.

Table 4. Case study results.

Traslation results	
src	中欧班列去程首次在此进行班列宽轨和标轨的换装作业
ref	for the first time, the china-europe train will carry out the reloading operation of the board rail and the standard rail.
mt1	for the first time, the central european banlei will carry out the replacement of the banliewi-de rail and the standard rail .
mt2	for the first time, the central china-europe banlei will carry out the replacement of the banli-ewide rail and the standard rail .

Evaluation results			
system	baseline score	QE+CpSqe score	HTER
mt1	0.0897	0.0687	0.2272
mt2	0.0832(-7.25%)	0.0516(-24.89%)	0.1818(-19.98%)

our document-level QE corpus with 1992 sentences. We randomly select 50% sentences to delete or replace 20%–70% of the words and enhance the corpus up to 2,565 sentences. Word2Vec model are trained on 23GB Chinese-English monolingual corpus from Wikipedia and Sohu News. CCMT19 Chinese-English parallel corpus and FBIS Chinese-English corpus are used to train the inner-extractor.

5.3 Preferred Centers Extraction

In this experiment, we use a rule-based method as the baseline. In the rule-based method, Stanfordnlp is used for syntactic analysis. Noun subject, clausal subject, direct object, indirect object are chosen to be preferred centers. The setup of our model is presented in Appendix B.

The experiments results are shown in Table 2. Our semi-supervised training method train model for two iterations on both Chinese and English data. The recall and accuracy of Chinese Model 3 achieve 60.70% and 59.44% respectively. And English Model 3 achieve 63.61% recall and 61.08% accuracy. Both semi-supervised model significantly outperform the rule based model. The performance of each iteration is better than that of last iteration indicating that our proposed semi-supervised method can improve the performance of model. We choose the recall as metrics for the reáson that we want to obtain comprehensive preferred centers as far as possible.

5.4 QE Results

In this experiment, we use Transformer-based feature extractor-evaluator as baseline model. Compared with the baseline, our model introduce an inner-extractor. The setup of CpQE model is shown in Appendix C. The result of quality estimation model is shown in Table 3. The Pearson correlation coefficients measure the correlation between model score and HTER. In the sentence-level QE, the difference among the three models is about 0.01. In document-level

QE, our CpQE+CpSeq model achieve the best performance with 0.6035, outperform the baseline by 0.0499. The rule-based Cp extractor with only 40.43% recall but still improve the QE model, indicating that not only preferred centers can improve the documen-level QE, other information also plays a role in the QE model. When the recall of Cp extraction increase, the performance of QE model further improve, which show the effectiveness of preferred centers. In the sentence-level QE, according to the setting of text boundary feature acquisition, the proposed model can not get any hint of the preferred center, which is equivalent to no additional information, so the performance of the model is comparable to that of the baseline model.

5.5 Case Study

As shown in Table 4, we provide the example of CpQE model and baseline model on scoring translation in document-level QE.

In the given example, the word "中欧班列 (china-europe train)" has two meanings. The first one is "the train from China to Europe" and the other one is "the train in central Europe". Since the previous sentences of the same document have mentioned "the train tack from Chengdu, China to Europe", the word in this sentence should be translated into "the train from China to Europe". Unfortunately, the translation output to be evaluated, i.e. mt1, provides an incorrect translation where the word "China" is missed. To test whether our proposed document-level QE system is sensitive to such errors, we simply recover the missing word "China" while ignore other mistakes in mt1 and produce another output, namely mt2. Then we evaluated these two outputs using the baseline model and our model, respectively. Clearly, the evaluation results show that both models indicate the decline of the edition rate. The proportion of the reduction of our model is higher than that of the baseline model, which is consistent with the HTER value, as listed in the fourth column. This results imply that our proposed model is more sensitive to such problems.

6   Conclusion

This research focus on the document-level machine translation quality estimation. Concretely, based on the concept of Preferred Center in the Centering Theory and the evaluation method of local text fluency, we manually annotated a small-scale dataset for Preferred Center extraction. Then, we trained a model to extract Preferred Centers for given texts and combine the extracted Preferred Centers as context information into the Predictor-Estimator model to improve the performance of QE. Furthermore, we construct a document-level Chinese-English QE dataset to measure the performance of our document-level QE models.

Acknowledgements. The authors would like to thank the three anonymous reviewers for their comments on this paper. This research was supported in part by the National Natural Science Foundation of China under Grant Nos. 62076211, U1908216 and 61573294 and the Outstanding Achievement Late Fund of the State Language Commission of China under Grant WT135-38.

A   Appendix

Algorithm 1. Running process of outer-extractor

Input: mt mCp src sCp
Output: Emb f_{outer}
1: do
2:   for i in range(T) do
3:     $[f_1, f_2, f_3] = \frac{2}{len(mt)+len(src)}[\text{len(sCp[i]), len(mCp[i]), len(mCp[i]) - len(sCp[i])}]$
4:     if mt[i] is the begining of the document do
5:       Emb[i] = 0
6:       $f_{o}uter = [f_1, f_2, f_3, 0, 0, 0, 1, 1, 0, 1, 1, 0]$
7:       continue
8:     Emb[i] = [Word2Vec(sCp[i-1]), Word2Vec(mCp[i-1])]
9:     $[f_4, f_5, f_6] = \frac{2}{len(mt)+len(src)}[\text{len(sCp[i-1]), len(mCp[i-1]), len(mCp[i-1]) - len(sCp[i-1])}]$
10:    $[f_7, f_8, f_9] = [\text{similarity(sCp[i-1], src[i]), similarity(mCp[i-1], mt[i]), similarity(sCp[i-1],}$
   src[i])- similarity(mCp[i-1], mt[i])]
11:      if mt[i] is the end of the document do
12:        $f_{outer} = [f_1, f_2, f_3, f_4, f_5, f_6, f_7, f_8, f_9, 1, 1, 0]$
13:        continue
14:      else do
15:        $[f_{10}, f_{11}, f_{12}] = [\text{similarity(sCp[i-1], sCp[i]), similarity(mCp[i-1], mCp[i]), similarity}$
   (sCp[i-1], sCp[i]) - similarity(mCp[i-1], mCp[i])]
16:        $f_{outer} = [f_1, f_2, f_3, f_4, f_5, f_6, f_7, f_8, f_9, f_{10}, f_{11}, f_{12}]$
17: **return** Emb, f_{outer}

The input of the outer-extractor is translation sentences mt, the preferred centers of translation sentences mCp, source sentences src and the preferred centers of source sentences sCp. The output of the extractor are embeddings of preferred centers Emb and the sentence relation features f_{outer}. T is the number of sentences in the corpus.

B Appendix

Table 5. Parameter of Bert-BiLSTM-CRF model

Parameter	Value	Describe
batch size	8	Total batch size for training
lr	0.01	The initial learning rate
epoch	10	Total number of training epochs to perform
lstm_size	128	LSTM hidden size
lstm_layers	1	Total number of LSTM layers
optim	Adam	Optimizer type

For preferred center extraction model, we use BERT-Base-Chinese as Chinese pre-trained model and BERT-Base as English pre-trained model. Some hyper-parameters are fixed: decoder layers are 12, hidden size of Bert is 768, the number of heads in multi-head attention is 12. Other parameters are shown in Table 5.

C Appendix

Table 6. Hyper-parameters of baseline predictor

Name	Value	Describe
src vocab size	120000	Size of vocabulary in source language
trg vocab size	120000	Size of vocabulary in target language
hidden size	512	Hidden size of Transformer
layers	2	Numbers of encoders and decoders in Transformer
head nums	8	Number of heads in multi-head attention
dropout	0.1	–
epoch	7	–
batch size	128	–
learning rate	2.0	–
optim	Lazyadam	Optimizer

Table 7. Hyper-parameters of baseline estimator

Name	Value	Describe
src vocab size	120000	Size of vocabulary in source language
trg vocab size	120000	Size of vocabulary in target language
unit nums	128	Unit numbers of BiLSTM
layers	1	Layers of BiLSTM
dropout	0.1	–
epoch	7	–
batch size	128	–
learning rate	2.0	–
optim	Lazyadam	Optimizer

Our CpQE model integrate an outer-extractor compared with baseline model. Other parameters is same as the baseline model. The parameters of baseline is shown in Table 6 and Table 7. The dimension of Word2Vec in outer-extractor is 512.

References

1. Banerjee, S., Lavie, A.: Meteor: an automatic metric for MT evaluation with improved correlation with human judgments. In: Proceedings of the ACL Workshop on Intrinsic and Extrinsic Evaluation Measures for Machine Translation and/or Summarization, pp. 65–72 (2005)
2. Chen, Z., et al.: Improving machine translation quality estimation with neural network features. In: Proceedings of the Second Conference on Machine Translation, pp. 551–555 (2017)
3. Devlin, J., Chang, M.W., Lee, K., Toutanova, K.: BERT: pre-training of deep bidirectional transformers for language understanding. arXiv preprint arXiv:1810.04805 (2018)
4. Doddington, G.: Automatic evaluation of machine translation quality using n-gram co-occurrence statistics. In: Proceedings of the Second International Conference on Human Language Technology Research, pp. 138–145 (2002)
5. Fan, K., Wang, J., Li, B., Zhou, F., Chen, B., Si, L.: "bilingual expert" can find translation errors. In: Proceedings of the AAAI Conference on Artificial Intelligence, vol. 33, pp. 6367–6374 (2019)
6. Felice, M., Specia, L.: Linguistic features for quality estimation. In: Proceedings of the Seventh Workshop on Statistical Machine Translation, pp. 96–103 (2012)
7. Fonseca, E., Yankovskaya, L., Martins, A.F., Fishel, M., Federmann, C.: Findings of the WMT 2019 shared tasks on quality estimation. In: Proceedings of the Fourth Conference on Machine Translation (Volume 3: Shared Task Papers, Day 2), pp. 1–10 (2019)
8. Grosz, B., Joshi, A., Weinstein, S.: Providing a unified account of definite noun phrases in discourse. In: Proceedings of the 21st Annual Meeting of the Association for Computational Linguistics. Association for Computational Linguistics (1983)
9. Grosz, B.J., Joshi, A.K., Weinstein, S.: Centering: a framework for modelling the local coherence of discourse (1995)

10. Kim, H., Jung, H.Y., Kwon, H., Lee, J.H., Na, S.H.: Predictor-estimator: Neural quality estimation based on target word prediction for machine translation. ACM Trans. Asian Low-Resour. Lang. Inf. Process. (TALLIP) **17**(1), 1–22 (2017)

11. Kim, H., Lim, J.H., Kim, H.K., Na, S.H.: QE BERT: bilingual BERT using multi-task learning for neural quality estimation. In: Proceedings of the Fourth Conference on Machine Translation (Volume 3: Shared Task Papers, Day 2), pp. 85–89 (2019)

12. Lafferty, J., McCallum, A., Pereira, F.C.: Conditional random fields: probabilistic models for segmenting and labeling sequence data (2001)

13. Li, M., Xiang, Q., Chen, Z., Wang, M.: A unified neural network for quality estimation of machine translation. IEICE Trans. Inf. Syst. **101**(9), 2417–2421 (2018)

14. Martins, A.F., Junczys-Dowmunt, M., Kepler, F.N., Astudillo, R., Hokamp, C., Grundkiewicz, R.: Pushing the limits of translation quality estimation. Trans. Assoc. Computat. Linguist. **5**, 205–218 (2017)

15. Papineni, K., Roukos, S., Ward, T., Zhu, W.J.: Bleu: a method for automatic evaluation of machine translation. In: Proceedings of the 40th Annual Meeting of the Association for Computational Linguistics, pp. 311–318 (2002)

16. Patel, R.N., et al.: Translation quality estimation using recurrent neural network. arXiv preprint arXiv:1610.04841 (2016)

17. Rubino, R., de Souza, J., Foster, J., Specia, L.: Topic models for translation quality estimation for gisting purposes (2013)

18. Snover, M., Dorr, B., Schwartz, R., Micciulla, L., Makhoul, J.: A study of translation edit rate with targeted human annotation. In: Proceedings of Association for Machine Translation in the Americas, Cambridge, MA, vol. 200 (2006)

19. Specia, L., Paetzold, G., Scarton, C.: Multi-level translation quality prediction with quest++. In: Proceedings of ACL-IJCNLP 2015 System Demonstrations, pp. 115–120 (2015)

20. Tan, Z., et al.: THUMT: an open-source toolkit for neural machine translation. In: Proceedings of the 14th Conference of the Association for Machine Translation in the Americas (AMTA 2020), pp. 116–122 (2020)

21. Tong, Y., Zheng, J., Zhu, H., Chen, Y., Shi, X.: A document-level neural machine translation model with dynamic caching guided by theme-rheme information. In: Proceedings of the 28th International Conference on Computational Linguistics, pp. 4385–4395 (2020)

22. Vaswani, A., et al.: Attention is all you need. In: Advances in Neural Information Processing Systems, pp. 5998–6008 (2017)

23. Voita, E., Sennrich, R., Titov, I.: When a good translation is wrong in context: context-aware machine translation improves on deixis, ellipsis, and lexical cohesion. arXiv preprint arXiv:1905.05979 (2019)

24. Walker, M.A., Joshi, A.K., Prince, E.F.: Centering in naturally-occurring discourse: an overview. In: Centering in Discourse. Citeseer (1998)

25. Yang, S., Wang, Y., Chu, X.: A survey of deep learning techniques for neural machine translation. arXiv preprint arXiv:2002.07526 (2020)

26. Yuan, Y., Sharoff, S.: Sentence level human translation quality estimation with attention-based neural networks. arXiv preprint arXiv:2003.06381 (2020)

SAU'S Submission for CCMT 2021 Quality Estimation Task

Yanan Li, Na Ye$^{(\boxtimes)}$, and Dongfeng Cai

Human-Computer Intelligence Research Center, Shenyang Aerospace
University, Shenyang 110136, China

Abstract. This paper describes our submissions to CCMT 2021 quality estimation sentence-level task for both Chinese-to-English (ZH-EN) and English-to-Chinese (EN-ZH). In this task. We follow TransQuest framework which is based on cross-lingual transformers (XLM-R). In order to make the model pay more attention to key words, we use the attention mechanism and gate module to fuse the last hidden state and pooler output of XLM-R model to generate more accurate prediction. In addition, we use the Predictor-Estimator architecture model to integrate with our model to improve the results. Experiments show that this is a simple and effective ensemble method.

Keywords: Quality estimation · XLM-R · Ensemble

1 Introduction

In recent years, with the development of neural network, the quality of machine translation has been greatly improved. However, it is still a problem whether the translated text needs further post-editing, which needs to be solved by translation quality estimation. Quality Estimation (QE) aims to evaluate the quality of machine translation without reference translation, which saves a lot of manpower and time and is more in line with the actual requirements.

This paper introduces in detail our submission of sentence-level quality estimation task. The sentence-level task aims to predict the Human-targeted Translation Edit Rate (HTER) [1] of the machine translation output, which reflects the editing distance from the translation to the correct reference translation. QE system needs to predict the HTER value, that is, the editing error rate of the translation, which is a regression problem.

Traditional quality estimation methods use time-consuming and expensive artificial features to represent source sentences and machine translations. QuEst++ [2] is a method based on machine learning. Later, researchers began to apply neural networks to generated neural features automatically to quality estimation tasks. However, the scarce quality estimation data can not give full play to the role of neural network. In order to solve this problem, researchers try to transfer bilingual knowledge extracted from parallel corpora to quality estimation tasks. This kind of work usually adopts the Predictor-Estimator model proposed by Kim et al. [3]. Fan et al. [4] introduced a bidirectional Transformer for predictor to extract features, and used 4-dimensional mis-matching features. Besides,

© Springer Nature Singapore Pte Ltd. 2021
J. Su and R. Sennrich (Eds.): CCMT 2021, CCIS 1464, pp. 16–24, 2021.
https://doi.org/10.1007/978-981-16-7512-6_2

Wang et al. [5] used Transformer-DLCL in predictor. Recently, the emergence of pre-training model has swept the whole field of natural language processing, and more and more researchers have begun to use pre-training model in quality estimation tasks. Pre-training model has been widely used in predictor and combined with appropriate estimator [6, 7]. At the same time, the ensemble method has been proved to be very effective to improve the results [8, 9].

TransQuest [10] is shown to achieve state-of-the-art results outperforming current open-source quality estimation frameworks when trained on datasets from WMT, so we use it as baseline and improve it. In order to make a better prediction, we have improved the output of this model to predict the translation quality more effectively. In addition, we use the ensemble method to integrate the above two models, which is simple but effective.

2 Methods

In this section, we describe the methods used by our submitted system. We first introduce the basic model we use, and then introduce our improvement methods based on this model.

2.1 Basic Model

We chose TransQuest as our basic model. TransQuest uses cross-language transformers model XLM-R [11], which is different from the previous predictor-estimator framework, because it does not use parallel corpus. Therefore, this model reduces the burden of complex neural networks and the demand for computing resources. TransQuest won the first place in WMT 2020 DA task, and achieved state-of-the-art results in the current open-source quality estimation frameworks in WMT datasets. The authors implement two different architectures, and we chose the MonoTransQuest architecture. The input of this model is to separate the original text and the translated text by [SEP] token and input them into XLM-R model together. Besides, they used the output of the [CLS] token as the input of a softmax layer. XLM-R is a multi-language pre-training model proposed by Facebook, which uses 2.5TB CommonCrawl to filter data, and masked language model pre-trained on text in 100 languages, which obtains state-of-the-art performance on cross-lingual classification, sequence labeling and question answering.

Another basic model is QE Brain [12] which follows the predictor-estimator architectures. They use a bidirectional Transformer [13] for predictor and bidirectional LSTM [14] for estimator. QE Brain constructed mis-matching features, and only using this feature to make predictions can get good results. The model can be directly understood as that if the quality of the translated text is very high, the word prediction model based on conditional language model can accurately predict the current word based on the context of the original sentence and the target sentence. On the contrary, if the translation quality is not high, it is difficult for the model to accurately predict the current words based on the context.

2.2 Proposed Method

TransQuest model follows the standard method of XLM-R classification, and uses the tensor corresponding to the first token [CLS] of the last layer for classification. We want to make full use of the output of XLM-R. Therefore, we adopt the method of fully fusing the information of pooler_output and last_hidden_state, using the attention mechanism [15] and gate module [16].

When dealing with pooler_output, we use the same operation as CLS token of last_hidden_state. We use dropout layer, linear layer and tanh nonlinear function to deal with it. The model structure is shown in Fig. 1.

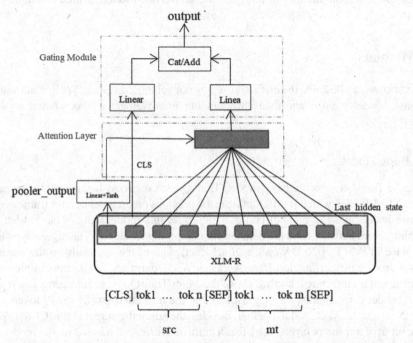

Fig. 1. Fusing pooler_output and last_hidden_state

2.2.1 Attention Mechanism

Considering that the method based on attention and weight has proved to be an effective way to selectively use additional information in many tasks, we add an attention layer. According to the contribution of words to tasks in different sentences, words are weighted to further enhance semantic information, and then pay attention to some keyword information. Different from the sequence-to-sequence task, the output of the translation quality estimation task is not a serialization process, so the attention weight of sentence vectors to word vectors in the current batch can be obtained only by one calculation, which is easy to implement.

We can simply regard the output of last_hidden_state as the word vector of the sequence, and the output of pooler_output as the sentence vector of the whole sequence. The formula for calculating attention is as follows:

$$\alpha_{i,j} = \frac{\exp(h_i \cdot e_{i,j})}{\sum_{j=i}^{m} \exp(h_i \cdot e_{i,j})} \tag{1}$$

$$v_i = \sum_{j=1}^{m} \alpha_{i,j} \, e_{i,j} \tag{2}$$

Where h_i denote the output of pooler_output, $e_{i,j}$ denote word embeddings in the output of last_hidden_state, and v_i is the output of the final attention layer.

2.2.2 Gate Module

Considering that the contribution of CLS token and attention vector in quality estimation task changes in different contexts, we hope to weight this information in the changing context through the gate module. We use a gate to control the information flow by

$$g = \sigma\left(W_1 \cdot h_{\text{cls}} + W_2 \cdot v_i + b_g\right) \tag{3}$$

$$u_i = \left[g \circ h_{\text{cls}}\right] + \left[(1-g) \circ v_i\right] \tag{4}$$

where W_1 and W_2 are trainable matrices and b_g the corresponding bias term. Then g is used to balance the information of CLS token and attention output, where h_{cls} denotes the CLS token, u_i denotes the output of the gate module and $\circ$ represents the element-wise multiplication operation. In the fusion mode of gate module, we try the addition and concatenation methods.

2.3 Ensemble

In order to further boost performance, we use the ensemble method. It is worth mentioning that we only use two models. One is the TransQuest model after our improvement, and the other is the QE Brain model. We chose QE Brain model because it works well on WMT data.

For the ensemble methods, due to time constraints, we only used the weighted average ensemble. According to the performance of the two models under CCMT data, we designed different weight ratios for ensemble, and finally chose the weight of the best result on the validation set for the ensemble experiment.

3 Experiment

3.1 Dataset

All the data we used came from CCMT2021, and no other extra data was used. The QE datasets have two language directions of both English-Chinese (EN-ZH) and Chinese-English (ZH-EN). The statistics of QE datasets are shown in Table 1. We don't use extra parallel data when using TransQuest framework, but we use QE Brain framework when we use ensemble methods, and the parallel corpus for training predictor comes from the machine translation task of CCMT2021. The statistics of parallel corpus are shown in Table 2.

Table 1. The statistics of QE datasets.

Direction	Aspect	Train	Dev	Test
EN-ZH	Sent	14,789	1,381	2,528
ZH-EN	Sent	10,070	1,143	2,412

Table 2. The statistics of parallel corpus.

Dataset	Data	Sentences
Datum2017	Train	999,985
Casict2015	Train	2,036,833
Casia2015	Train	1,050,000
Neu2017	Train	2,000,000
CCMT2019-en2zh	Dev	1,000

3.2 Settings

For sentence-level task, Pearson correlation coefficient is the main evaluation measure, In addition, we have set other measure: RMSE, MAE and Spearman Correlation.

We use the same settings for the two language directions pairs evaluated in this paper. We follow the default configuration of TransQuest framework, but adjust the learning rate to 2e−6, and other settings remain unchanged. We use Adam optimizer with a batch-size of eight. In the training process, the parameters of xlm-roberta-large model and the parameters of subsequent layers are updated. In the experiment, we used an NVIDIA Tesla T4 GPU.

We use different dropout rates for different language pairs. The final dropout rate is 0.4 in EN-ZH experiment and 0.3 in ZH-EN experiment.

3.3 Results of the Single Model

The results of CCMT2021 dev2019 are shown in Table 3 and Table 4. In order to get comparative experiments, the effectiveness of attention layer and gate module in translation quality estimation is demonstrated.

It can be seen from the results that after adding the attention output of pool and last, the use of add mode under the gate module has been improved by 3.67% in EN-ZH experiment and 2.76% in ZH-EN experiment, and the effect of using cat mode is not obvious. In addition, if the gate module is not used, the effect of direct addition will decrease, which may be because the attention output affects the semantic vector representation of the whole [CLS]. Therefore, there must be a gate module to control the attention output. If attention layer is not used, the promotion is not obvious, which fully proves that attention layer and gate model are indispensable.

Table 3. Results of the CCMT 2021 EN-ZH dev2019.

Model	Pearson	RMSE	MAE	Spearman
Baseline	0.5063	0.1552	0.1083	0.4439
+attention+cat	0.5069	0.1633	0.1104	0.4146
+attention+add	0.4835	0.1667	0.1152	0.3984
no_attention+gate_cat	0.5046	0.1628	0.1139	0.4328
no_attention+gate_add	0.5099	0.1620	0.1143	0.4409
+attention+gate_cat	0.5097	0.1633	0.1145	0.4355
+attention+gate_add	**0.5249**	0.1526	0.1081	0.4499

Table 4. Results of the CCMT 2021 ZH-EN dev2019.

Model	Pearson	RMSE	MAE	Spearman
Baseline	0.5204	0.1526	0.1070	0.4506
+attention+cat	0.5131	0.1513	0.1141	0.4527
+attention+add	0.4989	0.1654	0.1149	0.4376
no_attention+gate_cat	0.5211	0.1611	0.1130	0.4628
no_attention+gate_add	0.5215	0.1609	0.1124	0.4633
+attention+gate_cat	0.5228	0.1615	0.1104	0.4614
+attention+gate_add	**0.5348**	0.1501	0.1012	0.4927

3.4 Results of the Ensemble Methods

Through the experiment of different proportions of fusion, we obtained best results when we used the weights 0.7:0.3 in EN-ZH task and the weights 0.6:0.4 in ZH-EN task. The results of ensemble model are shown in Table 5.

According to the ensemble results, our improved TransQuest model has been improved by 6.4% in EN-ZH experiment and 7.8% in ZH-EN experiment, which fully reflects the effectiveness of the ensemble models. It can be concluded that integration is indeed a good way to improve the prediction accuracy in translation quality estimation.

A good model and a relatively bad model will produce better results. This is an interesting phenomenon, and perhaps it is also a question worth considering. QE Brain

Table 5. The results of ensemble model.

Model	EN-ZH	ZH-EN
Our improved best TransQuest	0.5249	0.5348
QE brain	0.3995	0.4639
ensemble	**0.5587**	**0.5765**

has been performing well under our previous WMT QE datasets, but its effect is not good under CCMT QE datasets, probably because the quality of CCMT translation is relatively good and involves a wide range of fields. The strength of the pre-training model is unexplainable to some extent, and the mis-matching feature is a feature that we think is very reasonable, so we will design some features more deeply in the future.

Although the prediction effect of QE Brain was not good on CCMT, after our analysis, we found that it could complement the prediction of TransQuest. We selected the first 400 examples of CCMT validation set to draw a line chart. As shown in the Fig. 2, we can see, the predictions of TransQuest are mostly distributed above the golden HTER value, while the predictions of QE Brain are mostly distributed below the golden HTER value. Therefore, the fusion of the two models can improve the prediction results of the ensemble methods.

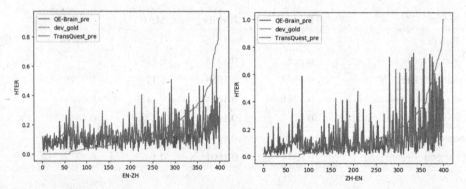

Fig. 2. The yellow line represents dev golden HTER, the blue line represents the predicted value of QE Brain and the green line represents the predicted value of our improved TransQuest. The left picture shows EN-ZH experiment, and the right picture shows ZH-EN experiment (Color figure online).

4 Conclusion

We describe our submissions to CCMT2021 QE sentence-level task. Our systems are based on TransQuest architecture and use QE Brain to make ensemble experiments. In order to make full use of the output of XLM-R and pay more attention to some key words, we use attention mechanism and gate module to fuse the output of XLM-R about last_hidden_state and pooler_output. Experiments show that this is effective. In addition, we also try to split last_hidden_state and add some external knowledge, but the effect is not good. On the other hand, the ensemble method is very effective in the task of translation quality estimation.

In the future work, although the pre-training model represents the source information and the target information in the same feature space, the source information is completely correct, while the target information contains wrong information. How to link the two more effectively is our next work. We want to introduce some external features to further enhance the performance. And we will also try some other ensemble methods in later experiments.

Acknowledgements. This work is supported by the Humanity and Social Science Foundation for the Youth researchers of Ministry of Education of China (19YJC740107), the National Natural Science Foundation of China (U1908216) and the Key Research and Development Plan of Liaoning Province (2019JH2/10100020).

References

1. Snover, M., Dorr, B., Schwartz, R., Micciulla, L., Makhoul, J.: A study of translation edit rate with targeted human annotation. In: Proceedings of Association for Machine Translation in the Americas, pp. 223–231. AMTA, Cambridge (2006)
2. Specia, L., Paetzold, G., Scarton, C.: Multi-level translation quality prediction with QuEst++. In: Proceedings of ACL-IJCNLP 2015 System Demonstrations, pp.115–120. ACL-IJCNLP, Beijing (2015)
3. Kim, H., Jung, H.Y., Kwon, H., Lee, J.H., Na, S.H.: Predictor-estimator: neural quality estimation based on target word prediction for machine translation. ACM Trans. Asian Low-Resour. Lang. Inf. Process. **17**(1), 1–22 (2017)
4. Fan, K., Wang, J., Li, B., Zhou, F., Chen, B., Si, L.: "Bilingual Expert" can find translation errors. In: Proceedings of the AAAI Conference on Artificial Intelligence, vol. 33, pp. 6367–6374 (2019)
5. Wang, Z., et al.: NiuTrans submission for CCMT19 quality estimation task. In: Huang, S., Knight, K. (eds.) CCMT 2019. CCIS, vol. 1104, pp. 82–92. Springer, Singapore (2019). https://doi.org/10.1007/978-981-15-1721-1_9
6. Wang, Z., et al.: Tencent submissions for the CCMT 2020 quality estimation task. In: Li, J., Way, A. (eds.) CCMT 2020. CCIS, vol. 1328, pp. 123–131. Springer, Singapore (2020). https://doi.org/10.1007/978-981-33-6162-1_12
7. Kim, H., Lim, J.H., Kim, H.K., Na, S.H.: QE BERT: bilingual BERT using multi-task learning for neural quality estimation. In: Proceedings of the Fourth Conference on Machine Translation (Volume 3: Shared Task Papers, Day 2), pp. 85–89, August 2019
8. Hu, C., et al.: The NiuTrans system for the WMT20 quality estimation shared task. In: Proceedings of the Fifth Conference on Machine Translation, pp. 1018–1023, November 2020
9. Cui, Q., Geng, X., Huang, S., Chen, J.: NJUNLP's submission for CCMT20 quality estimation task. In: Li, J., Way, A. (eds.) Machine Translation: 16th China Conference, CCMT 2020, pp. 114–122. Springer, Singapore (2020). https://doi.org/10.1007/978-981-33-6162-1_11
10. Ranasinghe, T., Orasan, C., Mitkov, R.: TransQuest: translation quality estimation with cross-lingual transformers. arXiv preprint arXiv:2011.01536 (2020)

11. Conneau, A., Khandelwal, K., Goyal, N., et al.: Unsupervised cross-lingual representation learning at scale. In: Proceedings of the 58th Annual Meeting of the Association for Computational Linguistics, pp. 8440–8451 (2020)
12. Fan, K., Wang, J., Li, B., et al.: "Bilingual Expert" can find translation errors. In: Proceedings of the AAAI Conference on Artificial Intelligence, vol. 33, pp. 6367–6374 (2019)
13. Devlin, J., Chang, M.-W., Lee, K., Toutanova, K.: BERT: pre-training of deep bidirectional transformers for language understanding. arXiv preprint, arXiv:1810.04805 (2018)
14. Graves, A., Schmidhuber, J.: Framewise phoneme classification with bidirectional LSTM and other neural network architectures. Neural Networks (2005)
15. Luong, M.T., Pham, H., Manning, C.D.: Effective approaches to attention-based neural machine translation. arXiv preprint arXiv:1508.04025 (2015)
16. Nie, Y., Tian, Y., Wan, X., et al.: Named entity recognition for social media texts with semantic augmentation (2020)

BJTU-Toshiba's Submission to CCMT 2021 QE and APE Task

Hui Huang[1], Hui Di[2], Hongjie Ren[1], Kazushige Ouchi[2], Jiahua Guo[1], Hanming Wu[1], Jian Liu[1], Yufeng Chen[1], and Jin'an Xu[1(✉)]

[1] School of Computer and Information Technology, Beijing Jiaotong University, Beijing, China
{18112023,20125222,17301093,17271137, jianliu,chenyf,jaxu}@bjtu.edu.cn
[2] Research & Development Center, Toshiba (China) Co., Ltd., Beijing, China
dihui@toshiba.com.cn, kazushige.ouchi@toshiba.co.jp

Abstract. This paper presents the systems developed by Beijing Jiaotong University and Toshiba (China) Co., Ltd. for the CCMT 2021 quality estimation (QE) and automatic-post editing (APE) task. For QE task, we mainly rely on multiple pretrained language models, and propose a multi-phase pre-finetuning scheme, to adapt the pretrained models to the target domain and task. The pre-finetuning scheme consists of language-adaptative finetuning, domain-adaptative finetuning and task-adaptative finetuning. For APE task, we use BERT-initialized Transformer as the backbone model, and create different groups of synthetic data by different data augmentation methods, i.e. forward translation, round-trip translation and multi-source denoising autoencoder. Multi-model ensemble is adopted in both tasks. Experiment results on the development set show high accuracy on both QE and APE tasks, demonstrating the effectiveness of our proposed methods.

Keywords: Machine translation · Quality estimation · Automatic post-editing

1 Introduction

This paper presents the systems developed by Beijing Jiaotong University and Toshiba (China) Co., Ltd. for the CCMT 2021 quality estimation (QE) and automatic-post editing (APE) task. For QE, we participate in the sentence-level task of Chinese-English direction. For APE, we participate in the task of Chinese-English direction.

Machine translation quality estimation aims to evaluate the quality of machine translation automatically without golden reference [2]. The quality can be measured with different metrics, such as HTER (Human-targeted Edit Error) [18]. Machine translation automatic post-editing aims to fix recurrent errors

© Springer Nature Singapore Pte Ltd. 2021
J. Su and R. Sennrich (Eds.): CCMT 2021, CCIS 1464, pp. 25–38, 2021.
https://doi.org/10.1007/978-981-16-7512-6_3

made by a certain decoder given the source sentence, by learning from correction examples [4]. Both the two tasks serve as a post-processing procedure for machine translation (MT) and are inner-related.

Both tasks rely on human-annotated triplets. QE is trained with triplets of *src* (source sentence), *mt* (machine translated sentence) and *score* (human-assessed score), and APE is trained with triplets of *src*, *mt* and *pe* (post-edited sentence). Since both human-assessment and post-editing require professional translators to manually annotate *src-mt* pairs, both tasks are highly data-scarce with only 10k-20k training examples. How to train a accurate estimator or post-editor with limited data remains a challenge.

For QE task, our system mainly relies on multiple pretrained models, including four multilingual pretrained models, i.e. multilingual BERT [8], XLM [6], XLM-RoBERTa-base and XLM-RoBERTa-large [5], and one monolingual model, i.e. RoBERTa [16]. We propose a multi-phase pre-finetuning scheme, to adapt the pretrained model to the target domain and task. The pre-finetuning procedure includes language-adaptive finetuning (LAF), domain-adaptive finetuning (DAF) and task-adaptive finetuning (TAF). We also jointly train the sentence-level estimator with word-level QE task. Different models are ensembled to achieve further improvement.

For APE task, we choose BERT-initialized Transformer [7] as the back-bone model, which uses the pretrained BERT to initialize the parameters of both encoder and decoder. We create synthetic triplets from openly-available parallel data using different methods, i.e. forward translation [17], round-trip translation [12] and multi-source denoising autoencoder. We build the multi-source denoising autoencoder to restore the corrupted reference given the source text, and the restored reference is deemed as the synthetic *mt*. We apply domain-selection to the parallel data for creating synthetic data, and different models trained with different data are ensembled to achieve further improvement.

Experiments on the development set shows we obtain competitive results in both directions, verifying the effectiveness of our proposed method.

2   Chinese-English Sentence-Level Quality Estimation

2.1   Model Description

Given the data-scarcity nature of QE, we build our system based on multiple pretrained models. We mainly rely on four multilingual pretrained models, i.e. multilingual BERT (abbreviated as mBERT) [8], XLM [6], XLM-RoBERTa-base (abbreviated as XLM-R-base) and XLM-RoBERTa-large (abbreviated as XLM-R-large) [5]. All of these four models are based on multi-layer Transformer [22] architecture, and are pretrained on massive multilingual text with shared multilingual vocabulary, enabling them to transfer to downstream tasks with limited training data.

We concatenate *src* (source sentence) and *mt* (machine translated sentence) following the way pre-trained models treat sentence pairs, and then feed the sentence pair to the model. We try two different strategies to aggregate the sentence-level representation, the first one is to directly use the first hidden representation of the pretrained model, and the second one is to add a layer of RNN on the top of the model, to better leverage the global context information, as shown in Fig. 1.

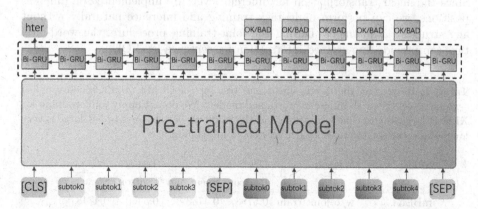

Fig. 1. Pretrained model for quality estimation with joint training. [**CLS**], [**SEP**] are predefined segment separators, and could be different in different models. The component circled with dashed line is alternative.

Although we mainly focus on sentence-level QE, the sentence and word-level QE are highly related, since their quality annotations are commonly based on the HTER measure [14]. During the calculation of sentence-level HTER score, the word-level QE tag for each word in mt could also be derived, and can serve as a supplementary information for training. Therefore, we implement multi-task learning, jointly train the sentence and word-level estimator together. The word-level estimation is based on the output logit according to each word, and we only use the logit of the first sub-token if one word is segmented into multiple sub-tokens. The loss function of both levels are defined as follow:

$$L_{word} = \sum_{s \in D} \sum_{x \in s} -(p_{ok} \log p_{ok} + \lambda p_{bad} \log p_{bad}),$$

$$L_{sent} = \sum_{s \in D} \| \, sigmoid(h(s)) - hter_s \, \|,$$

where s and x denote each sentence and word in the dataset D, and $h(s)$ is the hidden representation, and λ is a hyper parameter. Notice the quality of mt is very high [19], which means most of word-level tags are OK. To force the model to pay more attention to the erroneously translated words, we assign a weight λ for BAD words when calculating word-level loss. The loss of both sentence and word level are combined and back-propagated together, defined as follow:

$$L_{joint} = \sum_{s \in D} (L_{sent} + \eta \sum_{x \in s} L_{word}),$$

where η is a coefficient to balance the word-level and sentence-level loss. Since the linear transformation for different levels are implemented on different positions, we can perform multi-task training and inference naturally without any structure adjustment. During the joint-training procedure, the word-level tags can provide fine-grained information for sentence-level QE.

Table 1. Results on the development and test sets of CCMT 2021 Chinese-English sentence-level QE with different pretrained models. We do not apply joint training for XLM-R-large due to time limitation, and the result on dev set for XLM-R-large is very low because we set the max length very short in training.

Model	Method	Dev Set		Test set	
		Pearson	Spearman	Pearson	Spearman
mBERT	w/o joint train	0.5783	0.4768	0.5460	0.4748
	w/ joint train	0.5403↓	0.4339	0.5353↓	0.4254
XLM	w/o joint train	0.5464	0.4627	0.5368	0.4668
	w/ joint train	0.5388↓	0.4647	0.5335↓	0.4601
XLM-R-base	w/o joint train	0.5445	0.5077	0.4887	0.4443
	w/ joint train	0.5371↓	0.5143	0.4816↓	0.4388
XLM-R-large	w/o joint train	0.3643	0.3312	0.4736	0.4510

However, as shown in Table 1, joint training leads to degradation in all directions. This is not consistent with previous works which also apply joint training [11,15]. In the end, we decide to keep all the models for ensemble.

2.2 Multi-phase Pre-finetuning

Fine-tuning pre-trained language models on domain-relevant unlabeled data have become a common strategy to adapt the pretrained parameters to downstream tasks [9]. Previous works also demonstrate the necessity of pre-finetuning when performing QE on pretrained models [10,15]. In our system, we propose a multi-phase pre-finetuning scheme, consisting of language-adaptive finetuning (LAF), domain-adaptive finetuning (DAF), and task-adaptive finetuning (TAF). We pre-finetune the pretrained model on unsupervised parallel data with no quality annotations, by continuing performing mask language modeling.

LAF aims to adapt the pretrained model to bilingual concatenated pairs. Despite the shared multilingual vocabulary and training data, mBERT and XLM-R are originally monolingually trained, treating the input as either being from one language or another. But in our scenario, the input sentence pair is the

Table 2. Results on the development and test sets of CCMT 2021 Chinese-English sentense-leve QE. We do not apply LAF to XLM-R-large due to limited computation resource, and the result on dev set for XLM-R-large is very low because we set the max length very short in training.

Model	Method	Dev set		Test set	
		Pearson	Spearman	Pearson	Spearman
mBERT	Original	0.5783	0.4768	0.5460	0.4748
	+LAF	0.5875	0.4851	0.5547	0.4824
	+DAF	0.5933	0.4924	0.5589	0.4859
	+TAF	**0.5995**	0.5028	**0.5647**	0.4910
XLM	Original	0.5464	0.4627	0.5368	0.4668
	+DAF	0.5915	0.5065	0.5811	0.5053
	+TAF	**0.5942**	0.5304	**0.5838**	0.5077
XLM-R-base	Original	0.5445	0.5077	0.4887	0.4443
	+LAF	0.5699	0.5164	0.5110	0.4555
	+DAF	**0.5754**	0.5170	**0.5159**	0.4599
	+TAF	0.5716	0.5265	0.5103	0.4639
XLM-R-large	Original	**0.3643**	0.3312	0.4736	0.4510
	+DAF	0.3296	0.2996	0.5237	0.4961
	+TAF	0.2941	0.2674	**0.5379**	0.5090

concatenation of a bilingual parallel pair from two different languages. Therefore, we continue the mask language model on massive parallel sentence pairs (Table 2).

We use the parallel data from CCMT 2021 Chinese-English translation task, which contains roughly 9 million sentence pairs. We filter the data according to length and length ratio, and only keep sentence pairs with length shorter than 60, since we are unable to pre-finetune the pretrained model with *max_len* too big. The remaining 6 million pairs are used for LAF, which takes us roughly 10 days on two GPUs.

On the contrary, XLM is pretrained with the task of Translation Language Modeling, therefore we believe it is already adapted to bilingual concatenated sentence pair. Since LAF is performed on massive data with high computation overhead, we decide not to perform LAF on XLM.

DAF aims to adapt the pretrained model to the target domain. The representation of pretrained model is learned from the combination of various domains, and can be adapted to a certain domain if continued finetuning on unlabeled data from the domain. To this end, we select a domain-similar subset of the parallel data, and perform DAF for all the four pretrained models.

To be more specific, we finetune BERT as the domain classifier. The sentence pairs in the training and development set are deemed as in-domain data, and we randomly sample the same size of data as the general-domain data, for the training of classifier. We keep roughly 100k domain-similar sentence pairs for DAF, which takes us up to 3–4 hours on a single GPU.

TAF refers to pre-finetuning on the unlabeled training set for the given task. It uses a far smaller corpus (10k pairs) compared to DAF, but the data is much more task-relevant. We apply TAF for all the four models, and it is very fast with no more than 1 h on a single GPU.

The three-phase finetuninig scheme is performed in a pipelined manner, namely the latter phase is performed based on the parameters of the former phase. The representation of the pretrained model is adapted to our target language, domain and task, and can serve as a better start point to be finetuned on downstream task. Despite the limited training data, parallel data is readily accessible, therefore multi-phase finetuning is a convenient yet effective method to improve the performance without extra annotation.

2.3 Partial-Input Estimation

As denoted by Sun [20], QE systems trained on partial inputs perform as well as systems trained on the full input. Although the alignment information is absent, estimation can still be performed solely on the source text (to estimate the complexity) or solely on the target text (to estimate the fluency). This enables the incorporation of powerful monolingual models.

In our system, we perform partial-input estimation on the target side. We utilize the monolingual models of BERT and RoBERTa [16] to estimate the fluency. Only the target side of the bilingual pair is fed for training and evaluation. Despite the absence of the source text, the partial-input estimation still achieve high correlation because of the introduction of powerful monolingual model.

We also perform DAF and TAF to the monolingual model to adapt it to our scenario, as shown in Table 3.

Table 3. Results on the development and test sets of CCMT 2021 Chinese-English sentense-leve QE with partial-input.

Model	Method	Dev set		Test set	
		Pearson	Spearman	Pearson	Spearman
BERT-base	Original	0.5127	0.4652	0.4595	0.4177
RoBERTa-base	Original	0.5471	0.4656	0.4707	0.4279
RoBERTa-large	Original	0.5684	0.5133	0.4785	0.4350
	+DAF	**0.5715**	0.5407	**0.4903**	0.4457
	+TAF	0.5712	0.5063	0.4834	0.4395

2.4   Model Ensemble

After exhaustive hyper-parameter searching, we obtain more than ten strong models with different architectures and training procedures. To combine different predictions and achieve further improvement, we try two model ensemble techniques, namely averaging and linear regression. Averaging simply averages the predicted logits of different models. Linear regression learns a linear combination of different predictions using l_2-regularized regression over the dev set.

Table 4. Results on the development and test sets of CCMT 2021 Chinese-English sentense-leve QE. The results of single models are inconsistent with previous sections due to our final hyper-parameter searching.

Model	Dev set	Test set
	Pearson	Pearson
mBERT	0.6125	0.5581
XLM	0.6055	0.5800
XLM-R-base	0.5974	0.5454
XLM-R-large	0.2941	0.5379
RoBERTa	0.5681	0.4903
Averaging	0.6291	**0.6043**
Linear regression	**0.6376**	0.6034

As shown in Table 4, both two ensemble techniques achieve considerable improvement. Although the result of partial-input is comparatively low, it can provide complimentary information for other bilingual models when doing ensemble. Therefore, the incorporation of partial-input estimation is necessary.

3   Chinese-English Automatic Post-Editing

3.1   BERT-initialized Transformer

The current state of the art in APE is based on encoder-decoder structure with Transformer [22] as the backbone network. To alleviate the data-scarcity problem, we follow [7] and use multilingual BERT to initialize the parameters of Transformers, as shown in Fig. 2, which we call BERT-initialized Transformer. We follow their default setting, namely use the self-attention in BERT to initialize both the encoder and the decoder.

Specifically, instead of using multiple encoders to separately encode *src* and *mt*, we use BERT pre-training scheme, where the two strings after being concatenated by the **[SEP]** special symbol are fed to the single encoder, and assign different segment embeddings to each of them. Both the self-attention and context attention of the decoder are initialized with BERT. The self-attention and

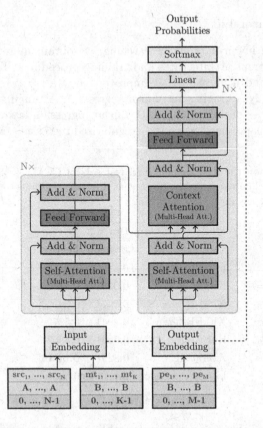

Fig. 2. BERT-initialized Transformer. Dashed lines show shared parameters.

Table 5. Results on the development set of CCMT 2021 Chinese-English APE with different architectures.

Model	Data	Dev set	
		TER	BLEU
Dual-source Transformer [13]	2 million	0.4585	41.74
Multi-source Transformer [21]	2 million	0.4344	46.39
BERT-based Transformer [7]	2 million	**0.4140**	**46.58**

embedding between encoder and decoder are shared, to reduce parameter size and improve training efficiency (Table 5).

We also compare with the dual-source transformer architecture of [13] and multi-source Transformer architecture of [21]. With 10k training triplets combined with 2 million synthetic triplets, the BERT-based Transformer outperforms the previous methods by a large margin, showing the effectiveness of pretrained parameters in APE task.

3.2   Domain Selection

Firstly we believe generative task is data-hungry, and therefore we use all the available parallel data to create synthetic triplets. We use the parallel data provided by CCMT 2021 Chinese-English translation, which consists of 23 million sentence pairs after filtering. However, during training we find that the model converges very soon and can not be improved afterwards. Therefore, we decide to apply domain selection for the synthetic data.

Table 6. Results on the development set of CCMT 2021 Chinese-English APE with different size of synthetic data. 10k refers to the model trained only with real data.

Model	Data	Dev Set	
		TER	BLEU
BERT-based Transformer	10k	0.4234	45.92
BERT-based Transformer	23 million	0.4679	39.57
BERT-based Transformer	5 million	0.4276	44.62
BERT-based Transformer	1 million	0.4089	47.80
BERT-based Transformer	200k	**0.4011**	**48.86**

To perform domain classification, we use the 10k training triplets as in-domain data, and randomly sample the same size of general domain data. We try two domain classification methods, [1] finetune BERT as a binary classifier, [2] use bilingual cross-entropy filtering method [1], and we use kenlm[1] to train 4-gram language models for filtering. Then synthetic triplets are combined with real triplets (which is oversampled 20 times) for training.

However, we do not see a clear difference between the two domain selection methods. On the contrary, we find that data size matters a lot. As shown in Table 6, we get the best result when incorporating 200k data. More data leads to domain irrelevance while only using the 10k real data is not enough for training. Therefore, we adopt the same data size in the following experiments.

3.3   Data Augmentation Techniques

Data augmentation is a de-factor paradigm for APE task [3]. The creation of synthetic data requires to generate synthetic *mt* given the parallel data (which are deemed as synthetic *src* and *pe*). Previous works rely on translation model to generate synthetic *mt* [12,17], but the connection between synthetic *mt-pe* is not consistent with real *mt-pe*. Actually, most synthetic *mt*s generated by machine translation are a correct translation of *src* but with different syntactic structure from *pe*. Forcing the APE model to transform the syntax of a correct translation is of little help to the training objective.

---

[1] https://kheafield.com/code/kenlm/.

In this work, we propose to generate synthetic *mt* via Multi-source Denoising Autoencoder (MDA), to better simulate the real error distribution. Denoising autoencoder is trained with two steps: (1) corrupt the text with an arbitrary noising function, and (2) learn a sequence-to-sequence model to reconstruct the original text. Specifically, in our scenario, we provide both the corrupted text and its corresponding translation to the encoder, leading to a multi-source denoising autoencoder structure, as shown in Fig. 3. The MDA learns to reconstruct the text based on its corruption and corresponding translation. This procedure is performed on massive publicly-available parallel sentence pairs (which are denoted as *src* and *ref*), without the need of extra annotations.

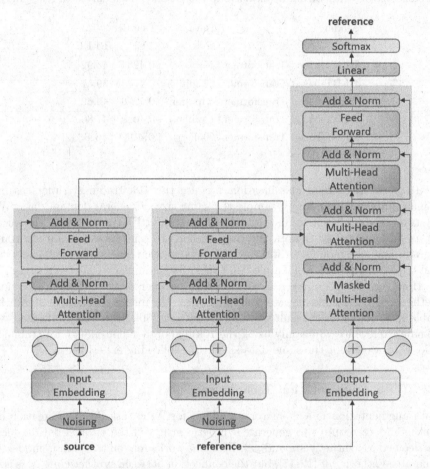

Fig. 3. Multi-source denoising autoencoder for generating synthetic triplets.

After that, the MDA can be used to generate synthetic triplets following the same formula. To be concrete, given parallel *src-ref* pairs, we would corrupt the *ref* by the same noising function, which is combined with *src* to generate reconstruction via MDA. Then the original and reconstructed *ref*s are deemed as *pe* and *mt*, respectively. The generated *mt* would inevitably differ *pe* (due to the corruption-reconstruction procedure), but their connection would be close since *mt* is inferred directly from *pe*. An also because the existence of source text, the restored *mt* would be not semantically far from the *src*. This is a better simulation of the MT error distribution.

Specifically, we try the combination of three noising transformations, i.e. word omission, word replacement and word permutation. Word omission randomly omits words in a sequence, and word replacement randomly replaces words, and word permutation randomly permutes words with a maximum distance. We use the 23 million CCMT 2021 Chinese-English data, and adopt two-fold jackknifing, namely split the data into two folds, one for training and another for decoding.

However, during the experiment, we find that if the corruption on the target side is too heavy, then the model would ignore the corrupted *pe* and only attend to the *src*. In that case, our multi-source denoising autoencoder would degrade to a normal machine translation model. Therefore, we try two strategies to force the model to attend to the corrupted target text.

[1] Corrupt the source text with similar flavor;

[2] Disturbing the embedding of the source text with Gaussian noise.

Both strategies make it difficult for the autoencoder to generate reference only relying on *src*, since the information of source side is also corrupted now. Therefore, it will try to restore the target sentence by both reorganising the corrupted *pe* and translating the disturbed *src*, leading to semantically deviated (but not unrelated), and syntactically consistent *mt*.

We also follow previous works and adopt forward translation and round-trip translation to create synthetic data. Forward translation [17] uses a forward-translation model to translate *src* to the target language as *mt*. Round-trip translation [12] uses two translation models, to translate *pe* firstly to the source language then to the target language, to generate synthetic *mt*. All the translation models are trained with the 23 million data with two-fold jackknifing.

Table 7. Results on the development set of CCMT 2021 Chinese-English APE with different augmentation methods. 200k synthetic triplets is combined with 10k real triplets oversampled 20 times.

Method	Noising		Dev set	
	Source	Reference	TER	BLEU
MDA	Gaussian	Corruption	0.4016	48.44
	Corruption	Corruption	0.4023	48.41
	None	Corruption	0.4035	48.39
Forward translation	–	–	0.4039	48.41
Round-trip translation	–	–	0.4011	48.86
Ensemble	–	–	**0.3953**	**49.20**

Although the MDA-based method does not outperform the round-trip translation based method, different methods lead to different data distributions and can provide complimentary information for each other. Therefore, we use all the models for ensemble, and achieve further improvement, as shown in Table 7.

4 Conclusion

In this paper, we described our submission in CCMT 2021 quality estimation and automatic post-editing task. For QE task, we verify that the pretrained models can be further improved on target language and target domain via prefinetuning, and incorporate powerful monolingual model to perform partial-input estimation. For APE task, we find that data-scarcity is alleviated to a large extent if use pretrained model to initialize the encoder-decoder, and propose to use multi-source denoising autoencoder to generate synthetic triplets.

Due to time limitation, we only participate in the Chinese-English direction. In the future, we will extend our system to QE and APE tasks on other languages, to verify the effectiveness of our proposed methods. Besides, we will also investigate how to combine these two inner-related tasks together to achieve further improvement.

Acknowledge. The research work descried in this paper has been supported by the National Key R&D Program of China 2020AAA0108001 and the National Nature Science Foundation of China (No. 61976015, 61976016, 61876198 and 61370130). The authors would like to thank the anonymous reviewers for their valuable comments and suggestions to improve this paper.

References

1. Axelrod, A., He, X., Gao, J.: Domain adaptation via pseudo in-domain data selection. In: Proceedings of the 2011 Conference on Empirical Methods in Natural Language Processing, Edinburgh, Scotland, UK, pp. 355–362 (2011)
2. Blatz, J., et al.: Confidence estimation for machine translation. In: COLING 2004: Proceedings of the 20th International Conference on Computational Linguistics, Geneva, Switzerland, 23–27 Aug 2004, pp. 315–321 (2004)
3. Chatterjee, R., Negri, M., Rubino, R., Turchi, M.: Findings of the WMT 2018 shared task on automatic post-editing. In: Proceedings of the Third Conference on Machine Translation, Volume 2: Shared Task Papers, Belgium, Brussels, pp. 723–738 (2018)
4. Chatterjee, R., Weller, M., Negri, M., Turchi, M.: Exploring the planet of the APEs: a comparative study of state-of-the-art methods for MT automatic post-editing. In: Proceedings of the 53rd Annual Meeting of the Association for Computational Linguistics and the 7th International Joint Conference on Natural Language Processing (Volume 2: Short Papers), Beijing, China, pp. 156–161 (2015)
5. Conneau, A., et al.: Unsupervised cross-lingual representation learning at scale. In: Proceedings of the 58th Annual Meeting of the Association for Computational Linguistics, Online, pp. 8440–8451 (2020)

6. Conneau, A., Lample, G.: Cross-lingual language model pretraining. In: Wallach, H., Larochelle, H., Beygelzimer, A., d'Alché-Buc, F., Fox, E., Garnett, R. (eds.) Advances in Neural Information Processing Systems, vol. 32 (2019)

7. Correia, G.M., Martins, A.F.T.: A simple and effective approach to automatic post-editing with transfer learning. In: Proceedings of the 57th Annual Meeting of the Association for Computational Linguistics, Florence, Italy, pp. 3050–3056 (2019)

8. Devlin, J., Chang, M.W., Lee, K., Toutanova, K.: BERT: pre-training of deep bidirectional transformers for language understanding. In: Proceedings of the 2019 Conference of the North American Chapter of the Association for Computational Linguistics: Human Language Technologies, Volume 1 (Long and Short Papers), pp. 4171–4186 (2019)

9. Gururangan, S., Marasović, A., Swayamdipta, S., Lo, K., Beltagy, I., Downey, D., Smith, N.A.: Don't stop pretraining: Adapt language models to domains and tasks. In: Proceedings of the 58th Annual Meeting of the Association for Computational Linguistics, Online, pp. 8342–8360 (2020)

10. Hu, C., et al.: The NiuTrans system for the WMT20 quality estimation shared task. In: Proceedings of the Fifth Conference on Machine Translation (2020)

11. Huang, H., Xu, J., Zhu, W., Chen, Y., Dang, R.: BJTU's submission to CCMT 2020 quality estimation task. In: Li, J., Way, A. (eds.) CCMT 2020. CCIS, vol. 1328, pp. 105–113. Springer, Singapore (2020). https://doi.org/10.1007/978-981-33-6162-1_10

12. Junczys-Dowmunt, M., Grundkiewicz, R.: Log-linear combinations of monolingual and bilingual neural machine translation models for automatic post-editing. In: Proceedings of the First Conference on Machine Translation: Volume 2, Shared Task Papers, Berlin, Germany, pp. 751–758 (2016)

13. Junczys-Dowmunt, M., Grundkiewicz, R.: MS-UEdin submission to the WMT2018 APE shared task: dual-source transformer for automatic post-editing. In: Proceedings of the Third Conference on Machine Translation: Shared Task Papers, Belgium, Brussels, pp. 822–826 (2018)

14. Kim, H., Lee, J.H., Na, S.H.: Predictor-estimator using multilevel task learning with stack propagation for neural quality estimation. In: Proceedings of the Second Conference on Machine Translation, Copenhagen, Denmark, pp. 562–568 (2017)

15. Kim, H., Lim, J.H., Kim, H.K., Na, S.H.: QE BERT: bilingual BERT using multi-task learning for neural quality estimation. In: Proceedings of the Fourth Conference on Machine Translation (Volume 3: Shared Task Papers, Day 2), Florence, Italy, pp. 85–89 (2019)

16. Liu, Y., et al.: Roberta: a robustly optimized BERT pretraining approach. CoRR abs/1907.11692 (2019). http://arxiv.org/abs/1907.11692

17. Negri, M., Turchi, M., Chatterjee, R., Bertoldi, N.: Escape: a large-scale synthetic corpus for automatic post-editing. In: Proceedings of the Eleventh International Conference on Language Resources and Evaluation (LREC 2018) (2018)

18. Snover, M., Dorr, B., Schwartz, R., Micciulla, L., Makhoul, J.: A study of translation edit rate with targeted human annotation. In: Proceedings of the 7th Conference of the Association for Machine Translation in the Americas: Technical Papers, pp. 223–231. Association for Machine Translation in the Americas, Cambridge (2006)

19. Specia, L., Blain, F., Logacheva, V., F. Astudillo, R., Martins, A.F.T.: Findings of the WMT 2018 shared task on quality estimation. In: Proceedings of the Third Conference on Machine Translation: Shared Task Papers, Belgium, Brussels, pp. 689–709 (2018)

20. Sun, S., Guzmán, F., Specia, L.: Are we estimating or guesstimating translation quality? In: Proceedings of the 58th Annual Meeting of the Association for Computational Linguistics, Online, pp. 6262–6267 (2020)
21. Tebbifakhr, A., Agrawal, R., Negri, M., Turchi, M.: Multi-source transformer with combined losses for automatic post editing. In: Proceedings of the Third Conference on Machine Translation: Shared Task Papers, Belgium, Brussels, pp. 846–852 (2018)
22. Vaswani, A., et al.: Attention is all you need. In: Guyon, I., Luxburg, U.V., Bengio, S., Wallach, H., Fergus, R., Vishwanathan, S., Garnett, R. (eds.) Advances in Neural Information Processing Systems, vol. 30 (2017)

Low-Resource Neural Machine Translation Based on Improved Reptile Meta-learning Method

Nier Wu, Hongxu Hou[✉], Xiaoning Jia, Xin Chang, and Haoran Li

College of Computer Science-college of Software,
Inner Mongolia University, Hohhot, China
cshhx@imu.edu.cn

Abstract. Multilingual transfer learning has been proved an effective method to solve the problem of low-resource neural machine translation (NMT). However, the global optimal parameters obtained through transfer learning can not effectively adapt to new tasks, which means the problem of local optimum will be caused when training the new task model. Although this problem can be alleviated by optimization-based meta-learning methods, but meta-parameters are determined by the second-order gradient term corresponding to the model parameters of a specific task, which consumes a lot of computing resources. Therefore, we proposed improved reptile meta-learning method. First, a multilingual unified word embedding method is proposed to represent multilingual knowledge. Secondly, the direction of meta-gradient is guided by calculating cumulative gradients on multiple specific tasks. In addition, the midpoint is taken as the meta-parameter in the space of the initial meta-parameter and the final task-specific model parameter to ensure that the meta-model has better multi-feature generalization ability. We conducted experiments in the CCMT2019 Mongolian-Chinese (Mo-Zh), Uyghur-Chinese (Uy-Zh) and Tibetan-Chinese (Ti-Zh), and the results show that our method has significantly improved the translation quality compared with the traditional methods.

Keywords: Meta-learning · Low-resource · Machine translation

1 Introduction

Low-resource NMT model is easy to produce over-fitting during model training due to the sparse data. In order to solve the problem of insufficient training sets for low-resource machine translation, there are two common methods: one is unsupervised learning [1], which uses large-scale monolingual corpus as an aid, expands pseudo-corpus by back translation or denoising self-encoding, and trains the model through self-learning or adversarial learning. However, the common pseudo-corpus noise reduction methods (deletion, replacement, addition) and shared word embedding mapping methods cannot fundamentally improve the

© Springer Nature Singapore Pte Ltd. 2021
J. Su and R. Sennrich (Eds.): CCMT 2021, CCIS 1464, pp. 39–50, 2021.
https://doi.org/10.1007/978-981-16-7512-6_4

noise and word alignment problem, so the current unsupervised machine translation effect is still lower than that of supervised model. Another method is transfer learning [2], which applies the model parameters learned from the high-resource language pair to the translation model of the low-resource language, and adapts the model to the low-resource task via fine-tuning. It mainly uses the prior knowledge of the high-resource language to assist the generation of the low-resource translation model [3].

Meta-learning is similar to transfer learning, which is essentially learning to learn. The meta-learning method is a model-independent method, which has better generalization ability and can quickly adapt to new tasks through a few training examples. Recently, there are mainly two methods for machine translation research using meta-learning: optimization-based method and model-based method. [4] proposed an optimization-based machine translation method for low-resource domains. Meta-parameters are iteratively learned through the proposed training strategies on translation tasks in different domains to adapt to translation tasks in new low-resource domains. [5] proposed an optimization-based meta-learning neural machine translation model training method. They used model-agnostic meta-learning (MAML) algorithm [6] to obtain shared initial parameters in multilingual large-scale language pairs, and the model can realize rapid convergence on low-resource translation tasks using initialized meta-parameters. Meanwhile, in order to solve the problem of inconsistency in word embedding space in multilingual translation tasks, the above studies all adopt a similar general word representation method [7] to adapt it to various meta-learning episodic.

Although the optimization-based meta-learning method shows potential in low-resource translation tasks, in the model training stage, the second-order gradient corresponding to the model parameters of a specific task will be repeatedly calculated, while consumes too much computing resources, and the performance of multi-task fitting is not ideal. Therefore, in order to avoid the above problems, we proposed an improved reptile meta-learning method. Specifically, it includes the following aspects.

- We proposed an unified word embedding representation method, which maps multiple languages including the target language into a new word embedding space instead of mapping to the word embedding space of the target language. This method improves the alignment accuracy between arbitrary languages without passing through the "pivot" language.
- We proposed an improved reptile meta-learning method, which can replace the original second-order gradient term to guide the direction of the meta-gradient, so that it has better multilingual knowledge transfer ability, and improves generalization performance while saving computing resources.

2   Background

Neural Machine Translation. Given the source language X, the neural machine translation model encodes X into a set of continuous intermediate representations, and the decoder decodes the target language Y from left-to-right according to the set of intermediate representations, as shown in Eq. 1.

$$p(Y|X;\theta) = \prod_{t=1}^{T+1} p(y_t|y_{0:t-1}, x_{1:T'};\theta) \qquad (1)$$

In general, recurrent neural network (RNN) is used to build the model. Recently, a decoder model with self-attention model and convolution structure has been proposed. Compared with the traditional model based on RNN method, the structure shows remarkable performance.

Low-Resource Machine Translation. Generally, unsupervised methods mainly include back translation [8] and dual learning [9]. While knowledge sharing methods mainly include transfer learning [10] and multi-task learning.

Meta-learning based NMT mainly draws lessons from MAML method, which includes two steps: meta-training and meta-testing. For meta-training, given a set of high-resource meta-translation tasks $(T1, ..., Tk)$, a set of tasks are sampled from the translation task generator each step, and the parameters are updated by MAML method to obtain the corresponding prior knowledge. For meta-testing, the low-resource translation model is initialized by using the learned parameters, so that the low-resource machine translation model can use prior knowledge and train a new translation model with a few number of samples. The learning process is shown in Eq. 2.

$$\theta^* = Learn(T^0; MetaLearn(T^1, ..., T^K)) \qquad (2)$$

For a specific low-resource language learning task T^0, the initial parameters are obtained from meta-model. It is assumed that the prior parameter distribution of the expected model satisfies isotropic gaussian distribution $N(\theta_i^0, 1/\beta)$. Meanwhile, to prevent the updated parameters from being far away from meta-parameters, the learning process of a specific language can be understood as maximizing logarithmic posteriori of model parameters for a given data set D_T, as shown in Eq. 3.

$$Learn(D_T, \theta^0) = \underset{\theta}{argmax} \sum_{(X,Y)\in D_T} logp(Y|X;\theta) - \beta \left\| \theta - \theta^0 \right\|^2 \qquad (3)$$

where X and Y represents the source language and target language of the data set, β is model parameter, $\left\| \theta - \theta^0 \right\|^2$ indicate modulo. In order to use high-resource language to repeatedly simulate low-resource translation episodic to obtain initialization parameters, the loss function of meta-learning is defined as Eq. 4.

$$Loss(\theta) = E_k E_{D_{T^k}, D'_{T^k}} \left[\sum_{(X,Y)\in D_{T^k}} logp(Y|X; Learn(D_{T^k}, \theta)) \right] \qquad (4)$$

3 Our Approach

We proposed a unified word embedding representation method, and an improved reptile meta-learning NMT method. As shown in Fig. 2.

3.1 Unified Word Embedding Representation

The vocabulary of each language only subject to an independent distribution space. To integrate multilingual knowledge, it is necessary to make universal representation of words in different languages. The common method is to map grammatically and semantically equivalent words from different languages to the same position in the vector space of the target language. Therefore, the mapping between other languages can be realized through the target language as a "pivot".

General methods such as cross-domain similarity local scaling (CSLS) optimize this mapping by minimizing the difference of word embedding of the same word in different languages. The optimal mapping matrix Q is constructed based on the loss of two norms, such as Eq. 5.

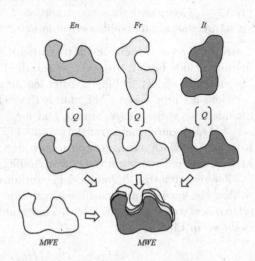

Fig. 1. Multi-aligned multilingual word embedding representation (MWE).

$$\min_{Q \in R^{d \times d}} \|XQ - PY\|_2^2 \tag{5}$$

where X is the mapped language, Y is the target language, and P represents the allocation matrix. However, this method needs bilingual dictionaries to assist and can only embed words in two languages. If multi-language embedding is done, only one language needs to be used as the transmission language. When there is no bilingual dictionary, $Wasserstein - Procrustes$ constraint is used on the allocation matrix P, so that the sum of each row and column of the allocation matrix is 1, and the matrix elements represent the degree of association of different words. Therefore, the allocation matrix P and the mapping matrix Q are optimized, and the 2-norm objective function embedded in multilingual words is obtained, such as Eq. 6.

$$\min_{Q \in Q_d, P \in P_n} \sum_i l(X_i Q_i, P_i X_0) \tag{6}$$

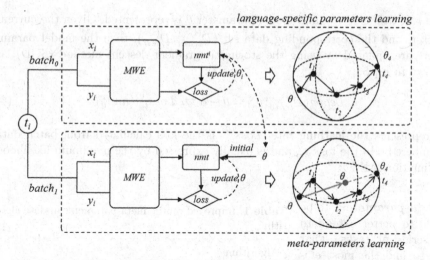

Fig. 2. In batch 0, the parameters of a specific task are learned and used it as initialization parameters for the next task. In batch 1, when the meta parameter is ready to be updated, there are two steps: 1. Utilize the cumulative gradient obtained by K-sampling as the direction of the meta-gradient. 2. The initial meta-parameters advance after K/2 steps and update (about half the distance between the final task-specific model parameter and the meta-parameters).

Among them, X_i, Q_i and P_i respectively represent the word embedding, mapping matrix and allocation matrix of the language mapping to the transfer language X_0. From Eq. 6, it can be seen that multilingual word embedding does not get the mapping between any two languages, but the mapping between one language and the target language (transfer language), which cannot guarantee the quality of word embedding except the target language. We use different cross-lingual word embedding methods to observed the quality of the translation. Therefore, we propose a new general vocabulary representation method: multi-aligned multilingual word embedding representation (MWE) As shown in Fig. 1. This is specifically shown in Eq. 7.

$$\min_{Q \in Q_d, P_{ij} \in P_n} \sum_{i,j} \alpha_{ij} l(X_i Q_i, P_{ij} X_j Q_j) \tag{7}$$

α represents weights, and i and j represent the number of languages. We take advantage of the fact that all word embedding maps to a unified space to realize better alignment.

3.2 NMT Method Based on Improved Reptile Meta-learning

Parameters of Task-Specific Model. Task-specific learning is similar to transfer learning. It mainly learns the model parameters of a specific tasks $\theta_\prime$ from high-resource translation tasks. Assume that the model corresponding to the i-th task is nmt_θ^i, and model parameter θ is represented. Given the current task t_i and the corresponding data set $(D_{train}^{(i)}, D_{test}^{(i)})$, then the model parameters are updated by using the stochastic gradient descent method (SGD), as shown in Eq. 8.

$$Learn(D_{train}^{(i)}; \theta') = \theta - \alpha \nabla_\theta Loss_{t_i}^{(0)}(nmt_\theta^i) \tag{8}$$

α represents the learning rate, $Loss_{t_i}^{(0)}$ is the loss calculated from batch data numbered 0 in the task t_i, and is usually expressed by the maximum likelihood estimation (MLE).

Meta-parameter. To be able to better extend to a series of tasks. That is, to find the most efficient parameter θ^* in the fine-tuning process after any given task, we need to re-sample a batch of data to update meta-parameters. If the corresponding loss function is set to $Loss_{t_i}^{(1)}$, the most efficient parameter θ^* and meta-parameter θ of the fine-tuning process expressed as shown in Eq. 9 and 10.

Table 1. Improved reptile meta-parameter update algorithm.

Algorithm
Require: $p(\tau)$: Distribution over tasks
Require: α, K:step hyper-parameters
Initialisation: Random θ
for $i = 1, 2, ..., n$ **do**
sample tasks $\tau_i \sim p(\tau)$
for all τ_i **do**
Evaluate the update $\theta_i = \theta - \alpha \nabla_\theta Loss_{\tau_i}(\theta)$
k times
end for
update:$\theta = \theta + \frac{2\alpha}{K} \sum_i^n (\theta_i - \theta)$
end for

$$\theta^* = \underset{\theta}{argmin} \sum_{t_i \sim p(t)} Loss_{t_i}^{(1)}(nmt_{\theta'}^i) \tag{9}$$

$$\theta \leftarrow \theta - \beta \nabla_\theta \sum_{t_i \sim p(t)} Loss_{t_i}^{(1)}(nmt_{\theta - \alpha \nabla_\theta Loss_{t_i}^{(0)}(nmt_\theta^i)}) \tag{10}$$

According to Eq. 10, after specific task parameters are learned in the inner loop, new data will be sampled from the same data set in the outer loop, and calculate a new gradient based on the same loss function, and meta parameters will be updated according to the new gradients. When sampling a new task, the initial meta-parameters are updated to the meta-parameters of the previous iteration, and the meta parameters are iteratively updated by repeatedly executing the previous steps. We proposed an improved reptile meta-learning method, when calculating the gradients of K tasks, we will no longer divide them into K different parameters θ_i^*, as shown in Table 1.

As shown in Table 1, given the initial parameter θ, k-round stochastic gradient descent of $SGD(Loss(nmt_\theta^i), \theta, k)$ is carried out according to $Loss(nmt_\theta^i)$, and then the parameter vector is returned. The version with batch samples multiple tasks at a single step. The gradient of our method is defined as $(\theta - W)/s$, where s is the step size used by SGD.

4 Experiments

4.1 Datasets

We use transfer learning based NMT (TF-NMT) and MetaNMT[1] as baselines and use 5 European (English (En), French (Fr), German (De), Spanish (Es), Italian (It)) and 3 Asian languages (Korean (Ko), Vietnamese (Vt), Japanese (Ja)) for the meta-training. All European language datasets from Europarl[2], however, all European languages use English as the target language instead of Chinese, so we adopt pivot-based method to construct an NMT model based on pivotal language (English) to obtain parallel sentence pairs from European languages to Chinese. Korean (Ko) corpus obtained from Korean Paral-

Table 2. The size of the training sample during meta-training and meta-test.

Corpus	sents.	src-tokens	trg-tokens
En-Zh	1.93M	3.22M	33.61
Fr-Zh	2.77M	51.39M	50.2M
De-Zh	1.92M	44.55M	47.81M
Es-Zh	1.96M	51.58M	49.09M
It-Zh	1.91M	47.4M	49.67M
Ko-Zh	0.54M	10.82M	11.11M
Vt-Zh	0.8M	15.95M	16.3M
Ja-Zh	0.68M	10.17M	11.23M
Mo-Zh	0.26M	8.85M	9.39M
Ti-Zh	0.4M	9.12M	8.68M
Ug-Zh	0.46M	10.12M	11.29M

lel Dataset,[3]. For Vietnamese (Vt), we use a crawler collect Vietnamese texts[4] from the Internet and then feed the Google translator[5] with the text, so that we obtain a loose parallel corpus between Vietnamese and Chinese. For Japanese (Ja), We conducted experiments with the ASPEC-JC corpus, which was constructed by manually translating Japanese scientific papers into Chinese. During meta-test period, we selected the following three different languages pairs (Mongolian-Chinese (Mo-Zh), Tibetan-Chinese (Ti-Zh), Uyghur-Chinese (Ug-Zh)) from CCMT2019: We use the officially provided train sets, valid sets, test sets for these languages. The size of the training sample is shown in Table 2.

---

[1] https://github.com/salesforce/nonauto-nmt.

[2] http://www.statmt.org/europarl.

[3] https://sites.google.com/site/koreanparalleldata.

[4] The Vietnamese corpus has 0.8 million Vietnamese sentences and 10 million Vietnamese monosyllables.

[5] https://translate.google.cn/?sl=vi&tl=zh-CN&op=translate.

4.2 Setting and Baseline

Setting. Our model is implemented using Pytorch[6], a flexible framework for neural networks. We base our model on the Transformer model and the released Pytorch implementation[7]. Parameters are set as follows: word embedding size = 300, hidden size = 512, number of layers = 4, number of heads = 6, dropout = 0.25, batch size = 128, and beam size = 5. Because our semantic space is

Table 3. Comparison of experimental results. Our model shows potential advantages in three different target tasks in a fully supervised environment.

Model	Mo-Zh	Ug-Zh	Ti-Zh
Transformer	28.15	23.42	24.35
TF-NMT	28.58	24.39	25.27
Meta-NMT	29.95	25.52	26.73
IR-Meta-NMT	30.83	26.29	27.18

obtained by multiple alignment of different languages. Therefore, we need to vectorize multilingual, here, two kinds of vectorization representation strategies are proposed: Static representation and dynamic representation. For static cross-lingual word embedding, we first employed FastText tools[8] to generate static monolingual word vector, and then use MUSE[9] or VECMAP[10] to generate cross-lingual representation. For dynamic cross-lingual word embedding, we obtained contextual dynamic word embedding through the ElMo model[11], and then use our multiple alignment approach[12] to get dynamic cross-lingual word embedding. In test phase, we use beam search to find the best translated sentences. Decoding ends when every beam gives an $\langle EOS \rangle$.

Baseline. We compared our approach against various baselines:

- Transformer[13]: The mainstream machine translation framework at this stage.
- TF-NMT[14]: A common method based on parameters transfer.
- Meta-NMT[15]: The method proposed by [5].
- IR-Meta-NMT: A model that improved reptile meta-learning methods that we proposed.

4.3 Result and Analysis

To observe the results, we give different experimental choices: First is to compare our model with a variety of experiments, mainly to observe the performance of

---

[6] https://pytorch.org/.
[7] https://github.com/pytorch/fairseq.
[8] https://github.com/facebookresearch/fastText.
[9] https://github.com/facebookresearch/MUSE.
[10] https://github.com/artetxem/vecmap.
[11] https://github.com/DancingSoul/ELMo.
[12] https://github.com/PythonOT/POT.
[13] https://github.com/tensorflow/tensor2tensor.
[14] https://github.com/ashwanitanwar/nmt-transfer-learning-xlm-r.
[15] https://github.com/MultiPath/MetaNMT.

Table 4. Low resource translation quality corresponding to various source datasets.

Meta-train	Mo-Zh		Ug-Zh		Ti-Zh	
	None	Finetune	None	Finetune	None	Finetune
Es It	9.98	14.61 ± .18	3.58	5.61 ± .18	4.41	4.51 ± .28
En Fr De	11.76	16.92 ± .3	4.05	7.25 ± .24	4.29	5.94 ± .15
European	14.53	19.08 ± .12	4.46	8.16 ± .08	5.17	6.91 ± .35
Ko	11.39	15.97 ± .25	6.39	10.38 ± .14	6.53	8.14 ± .16
Vt Ja	15.55	21.38 ± .11	7.11	9.57 ± .31	6.74	7.89 ± .15
Asia Languages	18.86	23.15 ± .29	10.76	11.41 ± .12	10.76	11.57 ± .10
All Languages	19.49	24.01 ± .27	11.12	12.56 ± .08	12.17	12.96 ± .19
Full Supervised	**31.76**		**27.1**		**28.35**	

our method. Second is the influence of different meta-learning datasets on the translation quality of the target tasks. As shown in Table 3 and 4. According to Table 3, we found that compared with the Transformer, BLEU scores of our method are increased by 2.68, 2.87 and 2.83 respectively in the three target tasks. In addition, compared with TF-NMT, the BLEU scores are also increased by 2.25, 1.9 and 1.91, which fully demonstrates that the global optimal parameters obtained from multilingual translation model training phase can not achieve better performance in low-resource translation tasks, because the gradient corresponding to the optimal parameter is easily introduced into the local minimum problem. Compared with [5], we also get same conclusions. They utilized conventional meta-learning algorithm and take the target language as the "pivot" to realized multilingual unified word representation. They query and locate the position of low-resource languages words in the unified semantic space via the key-value networks, and then integrated the multilingual knowledge. In addition, excessive consumption of computing resources during training phase. Therefore, our method has also been greatly improved in training efficiency, as shown in Table 5.

When we select different meta-training data, we found that the results are also different. When we select several large European languages, such as En, Fr, De, the parameters obtained by meta-

Table 5. Time consumption.

Model	Time consuming	Speedup
Meta-NMT	≈3day	–
IR-Meta-NMT	≈1.7day	1.76×

learning are transferred to the low-resource translation tasks, the BLEU scores is better than that of other European languages. However, the BLEU scores of the model is higher when Asia languages was used. In other words, whether model-dependent or model-independent methods are adopted, the effect will be further improved when there are some internal relations between the high-resource and low-resource languages, such as belonging to the same language family or having the same or similar grammatical structure.

4.4   Ablation Experiments

We observed the influence of various modules on the NMT model through ablation experiments, and analyzed the translation quality when using the CSLS, MWE, Meta-learning (ML)

Table 6. The ablation experiment.

Model	Mo-Zh		Ug-Zh		Ti-Zh	
	Dev	Test	Dev	Test	Dev	Test
ML+CSLS	28.78	28.16	24.2	23.35	25.77	25.28
ML+MWE	31.34	29.95	28.36	25.52	28.51	26.73
RML+CSLS	32.48	29.61	27.29	25.8	28.85	27.02
RML+MWE	**33.35**	**30.83**	**28.58**	**26.29**	**30.07**	**27.18**

and Reptile Meta-learning (RML). In addition, we also evaluated the impact of sentences of different lengths on the quality of the model. As shown in Table 6. We mainly use BPE to process the data. According Table 6, we found that when using the common meta-learning method, the NMT model represented by MWE word embedding is 1.79, 2.17 and 1.45 higher on the test set than the model represented by CSLS word embedding. It can be inferred that the MWE method has higher alignment accuracy and representation ability. Meanwhile, when using the improved reptile meta-learning method (RML), the translation quality of the MWE method is also better than that of the CSLS method. In addition, under the same conditions, the BLEU score of the test set using the RML method is increased by 0.88, 0.77 and 0.45 respectively compared with the model using the ML method, which is fully demonstrates the remarkable generalization ability of the model in this paper.

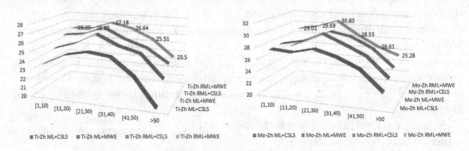

Fig. 3. The BLEU scores in different translation tasks.

According to the experiment shown in Fig. 3. The BLEU scores was highest when the sentence length was 20 to 30 words, and significantly decreased when the length was greater than 50 words.

4.5   Case Study

Case study include crosslingual word embedding alignment visualization and translation analysis. To observe the word embedding quality of meta training data and meta test data, we map Mongolian and English with the same semantics into Chinese vector space, aligned word pairs have more weight (cyan line), which proved that our method has a significant effect on crosslingual alignment, as shown in Fig. 4.

As shown in Fig. 5, we observe the translation results of different methods, and find that Transformer model has significant translation generation ability, which alleviates the problem of unknown words (UNK); TF-NMT and Meta-NMT methods ignore the relationship between source languages due to the problem of crosslingual word embedding mapping. Our method not only alleviates the above problems, but also learns more semantic representation including named entity (bold font), which shows remarkable effect.

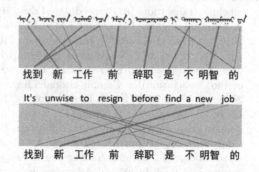

Fig. 4. Unified word embedding alignment visualization.

Source	كوردى تنزىلشىنى سەپكە قاراۋۇللىرىنىڭ ھۆرمەت سۆپىسىدىا پارت رەھبەرلىرى دۆلەت ئىككى.
Ref.	两国 元首 回 到 检阅台 观看 仪仗队 分列 式 。
Transformer	两个 国家 的 总统 去往 审查 台子 欣赏 分列 。
TF-NMT	两个 家园 主席 来到 检阅 <unk> <unk> 仪式 。
Meta-NMT	两国 首脑 来到 检阅 舞台 <unk> 队伍 仪式 散开 。
IR-Meta-NMT	**两国 元首** 返回 **检阅** 主席台 查看 队伍 **发散** 式 。

Fig. 5. Translation analysis.

5   Conclusion

In this paper, we proposed an improved reptile meta-learning method, in which the parameters of the previous specific task are taken as the initial parameters of the new specific task, and the final meta-parameter gradient is determined in combination with the first-order calculation method of the meta-gradient. Compared with the traditional method, this method is more efficient and effective. In addition, in order to integrate multi-language knowledge, we propose a multi-aligned cross-language word embedding, which alleviates the problems of knowledge sharing.

References

1. Jin, D., Jin, Z., Zhou, J.T., Szolovits, P.: Unsupervised domain adaptation for neural machine translation with iterative back translation. CoRR, vol. abs/2001.08140 (2020). https://arxiv.org/abs/2001.08140
2. Zoph, B., Yuret, D., May, J., Knight, K.: Transfer learning for low-resource neural machine translation. In: Proceedings of the 2016 Conference on Empirical Methods in Natural Language Processing, EMNLP 2016, Austin, Texas, USA, 1–4 November 2016, pp. 1568–1575 (2016). https://doi.org/10.18653/v1/d16-1163
3. Aji, A.F., Bogoychev, N., Heafield, K., Sennrich, R.: In neural machine translation, what does transfer learning transfer? In: Proceedings of the 58th Annual Meeting of the Association for Computational Linguistics, ACL 2020, Online, 5–10 July 2020, pp. 7701–7710 (2020). https://doi.org/10.18653/v1/2020.acl-main.688
4. Li, R., Wang, X., Yu, H.: MetaMT, a meta learning method leveraging multiple domain data for low resource machine translation. In: The Thirty-Fourth AAAI Conference on Artificial Intelligence, AAAI 2020, The Thirty-Second Innovative Applications of Artificial Intelligence Conference, IAAI 2020, The Tenth AAAI Symposium on Educational Advances in Artificial Intelligence, EAAI 2020, New York, NY, USA, 7–12 February 2020, pp. 8245–8252 (2020). https://aaai.org/ojs/index.php/AAAI/article/view/6339
5. Gu, J., Wang, Y., Chen, Y., Cho, K., Li, V.O.: Meta-learning for low-resource neural machine translation. In: Proceedings of the 2018 Conference on Empirical Methods in Natural Language Processing, Brussels, Belgium, October 31-November 4, 2018, pp. 3622–3631 (2018). https://doi.org/10.18653/v1/d18-1398
6. Finn, C., Abbeel, P., Levine, S.: Model-agnostic meta-learning for fast adaptation of deep networks. In: Proceedings of the 34th International Conference on Machine Learning, ICML 2017, Sydney, NSW, Australia, 6–11 August 2017, vol. 70. pp. 1126–1135 (2017). http://proceedings.mlr.press/v70/finn17a.html
7. Lample, G., Conneau, A., Ranzato, M.A., Denoyer, L., Jégou, H.: Word translation without parallel data. In: 6th International Conference on Learning Representations, ICLR 2018, Vancouver, BC, Canada, April 30 - May 3, 2018, Conference Track Proceedings (2018). https://openreview.net/forum?id=H196sainb
8. Abdulmumin, I., Galadanci, B.S., Isa, A.: Iterative batch back-translation for neural machine translation: a conceptual model. CoRR. vol. abs/2001.11327 (2020). https://arxiv.org/abs/2001.11327
9. He, D., et al.: Dual learning for machine translation. In: Advances in Neural Information Processing Systems 29: Annual Conference on Neural Information Processing Systems 2016, 5–10 December 2016, Barcelona, Spain, pp. 820–828 (2016). https://proceedings.neurips.cc/paper/2016/hash/5b69b9cb83065d403869739ae7f0995e-Abstract.html
10. Cheng, Y., Liu, Y., Yang, Q., Sun, M., Xu, W.: Neural machine translation with pivot languages. CoRR. vol. abs/1611.04928 (2016). http://arxiv.org/abs/1611.04928

Semantic Perception-Oriented Low-Resource Neural Machine Translation

Nier Wu, Hongxu Hou[✉], Haoran Li, Xin Chang, and Xiaoning Jia

College of Computer Science-college of Software, Inner Mongolia University,
Hohhot, China
cshhx@imu.edu.cn

Abstract. Pre-training method has been proved to significantly improve the performance of low-resource neural machine translation (NMT), while the common pre-training methods (BERT) uses attention mechanism based on Levenshtein distance (LD) to extract language features, which ignored syntax-related information. In this paper, we proposed a machine translation pre-training method with semantic perception which depend on the traditional position-based modeling, we uses semantic role labels (SRL) to annotate sentences with "predicate-argument" structures at the word level, and merge vectorized SRL with word vectors to deepen the model's understanding of deep semantics. In addition, to avoid parameter disaster, we proposed a hierarchical knowledge distillation method to fuse the NMT model and pre-training model to adapt to the output probability distribution of the pre-training model. We validated the method in the LDC En-Zh and CCMT2017 Mongolia-Chinese (Mo-Ch), Uyghur-Chinese (Uy-Ch), Tibetan-Chinese (Ti-Ch) tasks. The results show that compared with baseline, our model achieves significant results, which fully illustrates the generalization of the method.

Keywords: Pre-training · SRL · Machine translation

1 Introduction

The NMT method based on encoder-decoder framework [1,2] encodes the source language X into a set of continuous vector representation Z from left to right, and decodes the target language Y from Z in the same way. The model mainly adopts recurrent neural network (RNN) structure to encode the source language in a linear manner, resulting in the feature extraction ability being limited by the linear distance between the word embedding and its context, it means that correlation is inversely proportional to linear distance. In the early stage of training, there is less context information available and semantic relationships cannot be fully learned. Especially in low-resource tasks, the problem is more obvious due to sparse data. In order to alleviate these problems, [3] proposed a parallel encoding

© Springer Nature Singapore Pte Ltd. 2021
J. Su and R. Sennrich (Eds.): CCMT 2021, CCIS 1464, pp. 51–62, 2021.
https://doi.org/10.1007/978-981-16-7512-6_5

method, which is free from the constraints of time series encoding, and the understanding of language is no longer limited to its linear context information, but to extract features from the global perspective. While the Transformer model [4] based on the self-attention mechanism can learn the weight ratio between words in the language through the self-learning method, so that the model can improve the implicit learning ability of language knowledge. However, whether sequential encoding or parallel encoding is used, the model only learns the explicit structural information of the sentence (LD), and can not mine deeper grammatical information, resulting in the model effect not being significantly improved. Therefore, the NMT method that integrates grammatical information has gradually shows its potential. [5] proposed a syntactic attention-related NMT method to calculate the linear context and syntactic context information corresponding to the current target token by using the dual-attention mechanism, the decoder can accurately predict the target token according to the dual context representation. [6] proposed a novel NMT method for fusing abstract semantic representations (AMR), which used graph recurrent networks (GRN) to represent AMR information. Compared with syntactic tree, AMR greatly retains meaningful words in sentences and ignores non-contributing words, which improved the translation quality. [7] is similar to that of [6] in that it uses graph convolution network (GCN) to model the syntactic dependency tree corresponding to the source language and adds it to the top of the convolutional encoder to integrate word embedding and the syntax tree information.

For low-resource tasks, the quality of syntax-based NMT models is limited due to the lack of sufficient parallel corpus and corresponding syntax Treebank. In recent years, profit from the powerful semantic feature extraction ability of BERT, the pre-training method based on BERT has also been gradually applied to many NLP tasks including machine translation. [8] proposed a syntax-infused Transformer and BERT models for machine translation method, which adopted the BERT model to learn the position-aware context representation and regard it as the input of Transformer encoder. [9] proposed a BERT-based machine translation pre-training method, they fed the representation of BERT to all sublayers of NMT model and use the attention mechanism to adaptively control how each layer interacts with the representation of BERT. [10] proposed a knowledge distillation method using dynamic fusion strategy to provide pre-trained knowledge for NMT models. They use an adapter to dynamically transform the general representations in the pre-training model into representations more suitable for NMT model. In addition, the NMT model can fully learn the output distribution of the pre-training model through the knowledge distillation method. However, the pre-training method combined with rich semantic representation has not been widely applied to NMT tasks. With the advantages of BERT methods, this paper proposed a deep semantic perception assisted neural machine translation pre-training method, which includes the following contents.

- We proposed a target-oriented language pre-training method, which uses an improved BERT model to learn the implicit semantic features of the target language (the target language in this paper is Chinese).

- To alleviate the deficiency of syntax treebank, we adopted more easily available SRL labels to represent semantic information. We built a vector lookup table to obtain the vector corresponding to the SRL label, the use BiGRU to encode each vector, and finally splice different forms of SRL labels through the full connection layer.
- We use the hierarchical knowledge distillation method to instruct the output of each layer in the decoder, so that the NMT model can fully learn "prior knowledge" from the pre-training model.

2 Background

Neural Machine Translation. The NMT model based on attention mechanism simulates the translation probability $P(y|x)$ of the source language $X = \{x_1, ..., x_n\}$ to the target language $Y = \{y_1, ..., y_m\}$ word by word, as shown in Eq. 1.

$$P(Y|X) = \prod_{i=1}^{I} P(y_i|y_{<i}, X, \theta) \tag{1}$$

Where $y_{<i}$ indicates the partial translation result before the i-th decoding step, and θ indicates the parameters of the NMT model. The NMT model uses the maximum likelihood estimation method to optimize the parameters θ. For the parallel sentence pair $\{[x^n, y^n]\}_{n=1}^{N}$ in training set, the loss defined as shown in Eq. 2.

$$L_{CE} = \underset{\theta}{argmax} \sum_{n=1}^{N} log P(y^n|x^n; \theta) \tag{2}$$

Although the argumentation method is widely used, the problem of exposure deviation still exists, which also directly affects the quality of the NMT model.

NMT Assisted by Pre-training Method. The pre-training method transfers knowledge from resource-rich tasks to low-resource tasks. However, the NMT method takes the cross entropy between the two languages as the training goal to optimize the parameters, which is significantly different from the monolingual pre-training model.

Therefore, one approach is to use the resource-rich language pre-training model, and then put source language and the target language into the pre-training model to obtain the corresponding word embedding, and use pre-trained word embedding training NMT model. Another approach is to design a new sequence-to-sequence pre-training task to directly realize bilingual mapping in machine translation. Among them, XLM [11], MASS [12] and BART [13] are both cross-lingual pre-training method based on sequence-to-sequence.

3   Method

This section is mainly divided into the following aspects: semantic perception-assisted pre-training model and hierarchical knowledge distillation training process.

3.1   Semantic Perception-Assisted Pre-training Model

Obtain Semantic Role Label. SRL mainly takes the sentence as the unit and analyzes the predicate-argument structure of the sentence. Specifically, the task of SRL is to take the predicate as the center, explore the relationship between the various components in the sentence and the predicate, and use semantic roles to describe the relationship (argument). Generally, the process of SRL includes: syntax analysis-candidate argument pruning-argument recognition-argument labeling. According to the results of syntactic analysis, the part that is absolutely not an argument is pruned, and then the binary classification is used to determine whether the remaining part is argument, if so, the semantic category to which it belongs will be marked.

Given a sentence w, various predicate parameter structures are generated. To reveal the multidimensional semantics of sentences, we group different semantic labels corresponding to the same sentence and embed them with text into the next encoding component. The specific method is to input the sentence-predicate pairs (w, v) into the high-speed BiGRU to search for the predicate's argument, and the semantic role corresponding to argument is marked as y. The goal of prediction is to obtain the semantic role label sequence with the highest score among all possibilities Y. As shown in Eq. 3.

$$\hat{y} = \underset{y \in Y}{argmax} f(w, y) \tag{3}$$

Where $f(\cdot)$ indicate the nonlinear activation function in the BiGRU. To improve the prediction accuracy, the BIO constraints and semantic role label constraints are added. [14] made a detailed explanation, I won't repeat it.

Pre-training Model Combined with SRL. As shown in Fig. 1, pre-training model includes text encoding module and SRL encoding module. The text encoding module is similar to the common BERT. For sentence $X = \{x_1, ..., x_n\}$, we employ BERT model to capture the context information of each word segment and generate the corresponding context word embedding sequence.

For a sentence that contains the m semantic role label sequences associated with the predicate, $T = \{t_1, ..., t_m\}$, and the i-th label sequence t_i contains n labels, which can be expressed as $t_i = \{lbl_1^i, ..., lbl_n^i\}$, because semantic labels belong to the word level, the number of labels is equal to the sentence length. We construct a vector table that maps the semantic role labels to the corresponding vectors $\{v_1^i, ..., v_n^i\}$, and feed the vector to BiGRU to capture the hidden state

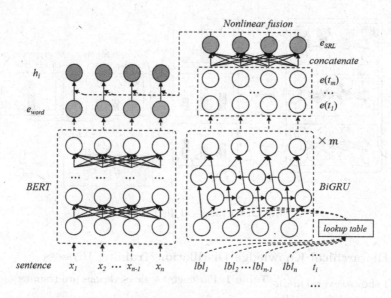

Fig. 1. Pre-training model combined with SRL. Word embedding (yellow circle) and semantic role representation (cyan circle) are obtained by BERT and BiGRU respectively. Then the new word embedding representation (purple circle) is obtained by nonlinear fusion method and fed to the NMT model. (Color figure online)

representation of the semantic labels, so as to extract the feature of the label sequences. As shown in Eq. 4.

$$e(t_i) = BiGRU(v_1^i, ..., v_n^i) \tag{4}$$

Where $0 < i \leq m$, we assume that L_i represents the set of label sequences corresponding to the i-th predicate token x_i, and the vector representation is defined as $e(L_i) = \{e(t_1), ..., e(t_m)\}$. Finally, we concatenate the m sequences of label representation and feed them to a fully connected layer to obtain an accurate label representation, as shown in Eq. 5.

$$\begin{aligned} e_{concat}(L_i) &= W[e(t_1), ..., e(t_m)] + b, \\ e_{SRL} &= \{e_{concat}(L_1), ..., e_{concat}(L_n)\} \end{aligned} \tag{5}$$

Where W indicates the weight, and b represents bias. e_{SRL} represents the embedding of the semantic label sequence corresponding to each predicate in a sentence.

The original sequence can be expressed as $e_{word} = \{e(x_1), ..., e(x_n)\}$. Then, word embedding and SRL embedding are concatenated by function $h = e_{SRL} \diamond e_{word}$.

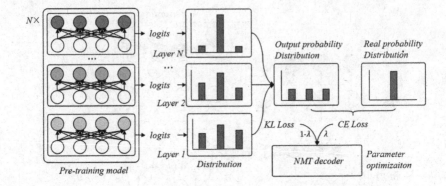

Fig. 2. Hierarchical distillation.

3.2   Hierarchical Knowledge Distillation Training Process

Table 1 shows several main-stream pre-training models at present. It can be seen that the parameters scale of the model is very

Table 1. Parameter scale of various pre-training models.

Model	ALBERT	BERT	GPT-2	XLNet	T5
Parameters	12mil	0.1bn	1.5bn	2bn	11bn

large, resulting in the method of parameter transfer or word embedding transfer method are difficult to achieve in neural machine translation model. Therefore, we proposed a hierarchical knowledge distillation method, which selectively extracts the output of a specific layer from the pre-training model and guides NMT model training according to its probability distribution. As shown in Fig. 2.

Generally, there is a temperature hyper-parameter τ in knowledge distillation method, and a smoother output distribution probability can be learned by increasing τ. The output probability is shown in Eq. 6.

$$p_{prt} = \frac{exp(z_{prt_i}/\tau)}{\sum_j exp(z_{prt_j}/\tau)} \tag{6}$$

Where z_{prt_i} represents hidden layer state, the equation also applies to the NMT model. Generally, a well pre-trained model will generate distributions with high probability for a few words, leaving others with probabilities close to zero. By increasing τ we expose extra information to the NMT model. In addition, for each layer of the pre-training model, it has certain representation ability. When some intermediate layers already have high distribution probability or confidence, the subsequent layers do not need to calculate KL divergence, which called "adaptive inference". While reducing resource consumption, it can still maintain a high prediction probability, so that the distillation data of the pre-training model has better indicative ability. The calculation of KL divergence is shown in Eq. 7.

$$D_{KL}(p_{prt}||p_{nmt}) = \sum_{i=1}^{N} p_{prt}(i) \cdot log\frac{p_{prt}(i)}{p_{nmt}(j)} \tag{7}$$

Where p_{prt} and p_{nmt} represent the probabilities of the pre-training model and NMT model respectively, the pre-training model is regarded as a teacher, NMT model is regarded as student. i represents the sentence number. Therefore, we define the KL divergence (relative entropy) loss between the output probability distribution of the pre-trained model and the NMT model, as shown in Eq. 8.

$$L_{KL}(p_{prt_0}, ..., p_{prt_{N-2}}, p_{nmt}) = \sum_{i=0}^{N-2} \tau^2 D_{KL}(p_{prt_i} \| p_{nmt}) \tag{8}$$

Where $p_{prt_0}, ..., p_{prt_{N-2}}$ represents the output probability of various intermediate layers in the pre-training model. Our goal is to minimize the KL divergence loss of the selected intermediate layers in the pre-training model and the NMT model, so that the two distributions are gradually similar.

To improve the utilization of the intermediate layers and ensure that the output probability distribution can optimize the NMT model, we set a threshold U, when the result is less than the threshold, we can distill the output data in advance. Otherwise, we need to calculate the output distribution of subsequent layers and repeat the process until the last layer of the pre-training model. The calculation of threshold U is shown in Eq. 9.

$$U = \frac{\sum_{i=1}^{N} p_{prt}(i) log p_{prt}(i)}{log \frac{1}{N}} \tag{9}$$

The variables in the Eq. 9 have been explained above and will not be repeated. For convenience, we use entropy as the threshold value. If the entropy value is large, the confidence is low, and if the entropy value is small, the result can be output, which not only saves subsequent calculation resources but also improves the inference speed. For this reason, the objective function of the model can be regarded as the weighted sum of cross-entropy loss (See Eq. 2) and relative entropy loss (See Eq. 8), as shown in Eq. 10.

$$L = \lambda L_{CE} + (1 - \lambda) L_{KL} \tag{10}$$

We set λ to 0.5.

4   Experiments

4.1   Datasets and Configuration

We conducted experiments on English-Chinese (En-Zh) and three low-resource translation tasks (Mo-Zh, Uy-Zh, Ti-Zh). For En-Zh task, the training set consist of 1.2 million bilingual sentences from LDC corpus[1], we use NIST02 as validation set, and NIST03-06 as the test set. For low-resource translation task, the data sets are provided by CCMT2017, as shown in Table 2.

---

[1] http://www.ldc.upenn.edu/.

In addition, we limit the bilingual vocabulary to 35K words and limit the length of sentences to 80. We utilized BLEU scores[2] to evaluate the quality. The parameters are updated by stochastic gradient descent (SGD), and the learning rate is dynamically adjusted by adam and the initial value is set to 0.0001. Word embedding dimension and hidden layer set 512, the beam size is 8, we apply dropout to avoid over-fitting, with dropout rate being 0.2, set U to 0.2. Since the pre-training model combined with SRL is used to guide the distribution probability of the target language prediction, we perform semantic role labeling on the target language (Chinese), and the label sets comes from Chinese Binzhou Proposition Bank (CPB)[3].

Table 2. Data sets for three low-resource machine translation tasks.

	Training	Valid	Test
Mo-Zh	64752	500	500
Uy-Zh	542796	1000	1000
Ti-Zh	30004	500	500

Our pre-training model is improved on SemBERT[4]. For the NMT model, we improved the Transformer[5] to implement our approach. To verify the effectiveness of the model, we also adopted two NMT methods combined with the pre-training model as the baseline,**BERT4NMT** [9]: A NMT method that integrated BERT pre-training model, and extracted the knowledge of pre-training model by introducing BERT-based attention mechanism.**AK4NMT** [10]: They used fusion strategy to transformed pre-trained word embedding into a more suitable representation for NMT task, and employed the distillation method to learn the output probability distribution of the pre trained model. We employed two TITAN X to train the model and obtained by averaging the last 5 checkpoints for the translation tasks.

4.2   Results and Analysis

Results. According to Table 3, our model compared to the traditional Transformer, BLEU scores improved by 1.75, 0.49, 1.81, 1.28 and 2.58, respectively, and the average BLEU scores increased by 1.18.

Table 3. En-Zh translation results in LDC corpus.

Model	NIST02	NIST03	NIST04	NIST05	NIST06	AVG
Transformer	37.19	36.66	37.06	34.89	35.23	35.96
BERT4NMT [9]	38.01	36.67	38.45	34.92	34.02	36.01
AK4NMT [10]	37.92	36.25	38.08	35.53	36.29	36.54
Our model	**38.94**	**37.15**	**38.87**	**36.17**	**37.81**	**37.5**

---

[2] https://github.com/moses-smt/mosesdecoder/blob/master/scripts/generic/multi-bleu.perl.

[3] http://verbs.colorado.edu/chinese/cpb/.

[4] https://github.com/cooelf/SemBERT.

[5] https://github.com/tensorflow/tensor2tensor.

Table 4 shows the experimental results of three low-resource translation tasks, it can be seen that our method has also improved 4.18, 3.61 and 3.59 BLEU scores in three low-resource machine translation tasks.

Analysis. Compared with two NMT models based on pre-training methods, our pre-training method combines SRL to obtain additional information. Meanwhile, due to the limitation of the MLE algorithm, the NMT model only relies on real translations to predict the target word,

Table 4. The BLEU scores for three low-resource machine translation tasks.

Model	Mo-Zh	Uy-Zh	Ti-Zh
Transformer	27.18	32.41	24.29
BERT4NMT [9]	29.03	34.96	25.85
AK4NMT [10]	29.51	35.77	26.09
Our Model	**31.36**	**36.02**	**27.88**

making the output distribution of the model more concentrated and cannot be effectively generalized to other words. Therefore, the distillation data output by the pre-trained model is better used to optimize the NMT model by adjusting the temperature hyper-parameter τ, and then extract more semantic representations.

4.3   Ablation Experiment

The ablation experiment in this paper is mainly used to observe the effect of the proposed method on the quality of the translation, including whether the pre-training model is combined with SRL, whether the encoder and decoder of the NMT model use pre-trained word embedding, and whether

Table 5. The ablation experiment.

Pre-training	Module	En-Zh	Mo-Zh	Uy-Zh	Ti-Zh
BERT	Emb-Enc	36.1	28.15	32.98	25.26
	Emb-All	36.02	28.28	33.05	25.47
	+KD	36.39	29.78	33.96	26.17
	+H-KD	36.92	30.11	34.79	26.85
SRLBERT	Emb-Enc	36.59	29.72	33.18	26.12
	Emb-All	36.72	29.26	33.25	25.87
	+KD	37.03	30.75	34.98	27.17
	+H-KD	**37.5**	**31.36**	**36.02**	**27.88**

knowledge distillation or hierarchical knowledge distillation is used. See Table 5 for details.

According to Table 5, when using general pre-training word embedding, whether the word embedding applied to the encoder (Emb-Enc) or encoder-decoder (Emb-All), it does not significantly improve the quality of translation. However, when the word embedding generated by the BERT model integrated with semantic role labels (SRLBERT) is used, the quality of the model has been improved to a certain extent. Meanwhile, compared with the NMT model using general knowledge distillation (KD), the hierarchical knowledge distillation (H-KD) method has better representation ability and translation effect. This shows that the intermediate representation of the pre-trained model also has accurate semantic feature extraction ability.

Fig. 3. BLEU scores for sentences of different length.

4.4 Case Study

We observed the BLEU scores with different length and the quality of translation obtained by various methods. According to Fig. 3, when the sentence length is about 20 to 30 tokens, the model has better performance. It can be seen that the NMT model using pre-training word embedding combined with semantic role labeling (SPTNMT) and hierarchical knowledge distillation (H-KD) methods has better generalization performance and translation fidelity.

	᠊᠊᠊᠊᠊᠊᠊᠊ ᠊᠊᠊ ᠊᠊᠊᠊᠊᠊ · ᠊᠊᠊ ᠊᠊᠊᠊ ᠊᠊ ᠊᠊᠊᠊᠊᠊᠊᠊᠊᠊ ᠊᠊ ᠊᠊᠊᠊ ᠊᠊᠊᠊᠊᠊/ · ᠊᠊᠊᠊᠊᠊᠊᠊ ᠊᠊ ᠊᠊᠊ ᠊᠊ ᠊᠊᠊᠊᠊᠊᠊/ ᠊ ᠊᠊᠊᠊᠊ ᠊᠊᠊ ᠊᠊ ᠊᠊᠊᠊᠊ ᠊᠊ ᠊᠊/ ··
Reference	被告 的 辩护 律师 想 把 案件 发生 时 被告 精神 失常 ， 为 被告 不 承担 法律 责任 进行 辩护 。
Transformer (NMT)	被告 辩护 律师 把 案件 被告 神经 失准 ， 为 不 承担 法律 责任 辩护 。
Pre-trainingNMT(PT-NMT)	被告 辩论 士 想 案件 发生 完毕 被告 失落 ，使 被告 不 承担 责任 辩护 。
SRL_PT-NMT	被告 辩论 律师 把 案件 发生 后 被告 精神 失落 ，使 被告 无 法 承担 法律 辩护 。
+KD	被告 的 辩论 律师 把 案件 发生 被告 精神 错乱 ，使 被告 无 法 承担 责任 进行 辩论 。
+H-KD	被告 的 辩护 律师 把 案件 发生 时 被告 精神失常 ，为了 被告 不用 承担 法律责任 辩解 。

Fig. 4. Case analysis.

As shown in Fig. 4, taking Mo-Zh translation tasks as an example, our method significantly improves the translation fluency and faithfulness compared with the translation generated by the pre-training model based on BERT. It can be seen that our method pays more attention to semantic coherence in the process of context generation. Meanwhile, due to the use of pre-trained word embedding combined with semantic roles labels, our proposed NMT model can effectively representation the context when predicting verbs or nouns and the words with reference relations. In addition, the hierarchical knowledge distillation method can also be used to provide more choices for translation, so as to significantly improve the generalization ability of the model.

5   Conclusion

We proposed a pre-training method that integrates semantic role labeling, and embed the words generated by the pre-training model into the NMT model for training. Meanwhile, to improve the prediction accuracy of the decoder and improve generalization ability, we proposed a hierarchical knowledge distillation method to guide the NMT model to learn the output probability distribution of the pre-trained model, so that the NMT model can comprehensively learn the probability distribution of the translation. Experiments show that our method has shown significant effects on large-scale corpus translation tasks and low-resource translation tasks. In the future, we will continue to study syntax-based pre-training methods and merge with NMT model to improve translation quality.

References

1. Sutskever, I., Vinyals, O., Le, Q.V.: Sequence to sequence learning with neural networks. In: Advances in Neural Information Processing Systems 27: Annual Conference on Neural Information Processing Systems, pp. 3104–3112 (2014). http:// papers.nips.cc/paper/5346-sequence-to-sequence-learning-with-neural-networks
2. Bahdanau, D., Cho, K., Bengio, Y.: Neural machine translation by jointly learning to align and translate. In: 3rd International Conference on Learning Representations, ICLR 2015, San Diego (2015). http://arxiv.org/abs/1409.0473
3. Gehring, J., Auli, M., Grangier, D., Yarats, D., Dauphin, Y.N.: Convolutional sequence to sequence learning. In: Proceedings of the 34th International Conference on Machine Learning, ICML 2017, Sydney, pp. 1243–1252 (2017). http:// proceedings.mlr.press/v70/gehring17a.html
4. Vaswani, A., et al.: Attention is all you need. In: Advances in Neural Information Processing Systems 30: Annual Conference on Neural Information Processing Systems, 4–9 December 2017, Long Beach, CA, USA, pp. 5998–6008 (2017). http:// papers.nips.cc/paper/7181-attention-is-all-you-need
5. Chen, K., Wang, R., Utiyama, M., Sumita, E., Zhao, T.: Syntax-directed attention for neural machine translation. In: Proceedings of the Thirty-Second AAAI Conference on Artificial Intelligence, (AAAI-18), the 30th innovative Applications of Artificial Intelligence (IAAI-18), and the 8th AAAI Symposium on Educational Advances in Artificial Intelligence (EAAI-18), New Orleans, Louisiana, USA, 2–7 February 2018, pp. 4792–4799 (2018). https://www.aaai.org/ocs/index.php/ AAAI/AAAI18/paper/view/16060
6. Song, L., Gildea, D., Zhang, Y., Wang, Z., Su, J.: Semantic neural machine translation using AMR. Trans. Assoc. Comput. Linguist. 7, 19–31 (2019). https:// transacl.org/ojs/index.php/tacl/article/view/1474
7. Marcheggiani, D., Bastings, J., Titov, I.: Exploiting semantics in neural machine translation with graph convolutional networks. In: Proceedings of the 2018 Conference of the North American Chapter of the Association for Computational Linguistics: Human Language Technologies, NAACL-HLT, New Orleans, Louisiana,

USA, 1–6 June 2018, Volume 2 (Short Papers), pp. 486–492 (2018). https://doi. org/10.18653/v1/n18-2078

8. Sundararaman, D., et al.: Syntax-infused transformer and BERT models for machine translation and natural language understanding. CoRR, vol. abs/1911.06156 (2019). http://arxiv.org/abs/1911.06156

9. Zhu, J., et al.: Incorporating BERT into neural machine translation. In: 8th International Conference on Learning Representations, ICLR 2020, Addis Ababa, Ethiopia, 26–30 April 2020 (2020). https://openreview.net/forum?id=Hyl7ygStwB.

10. Weng, R., Yu, H., Huang, S., Cheng, S., Luo, W.: Acquiring knowledge from pre-trained model to neural machine translation. In: The Thirty-Fourth AAAI Conference on Artificial Intelligence, AAAI 2020, The Thirty-Second Innovative Applications of Artificial Intelligence Conference, IAAI 2020, The Tenth AAAI Symposium on Educational Advances in Artificial Intelligence, EAAI 2020, New York, NY, USA, 7–12 February 2020, pp. 9266–9273 (2020). https://aaai.org/ojs/ index.php/AAAI/article/view/6465

11. Conneau, A., Lample, G.: Cross-lingual language model pretraining. In: Advances in Neural Information Processing Systems 32: Annual Conference on Neural Information Processing Systems 2019, NeurIPS 2019, 8–14 December 2019, Vancouver, BC, Canada, pp. 7057–7067 (2019). https://proceedings.neurips.cc/paper/2019/ hash/c04c19c2c2474dbf5f7ac4372c5b9af1-Abstract.html

12. Song, K., Tan, X., Qin, T., Lu, J., Liu, T.-Y.: MASS: masked sequence to sequence pre-training for language generation. In: Proceedings of the 36th International Conference on Machine Learning, ICML 2019, 9–15 June 2019, Long Beach, California, USA, vol. 97, pp. 5926–5936 (2019). http://proceedings.mlr.press/v97/song19d. html

13. Lewis, M., et al.: BART: denoising sequence-to-sequence pre-training for natural language generation, translation, and comprehension. In: Proceedings of the 58th Annual Meeting of the Association for Computational Linguistics, ACL 2020, Online, 5–10 July 2020, pp. 7871–7880 (2020). https://doi.org/10.18653/v1/2020. acl-main.703

14. He, S., Li, Z., Zhao, H.: Syntax-aware multilingual semantic role labeling. In: Proceedings of the 2019 Conference on Empirical Methods in Natural Language Processing and the 9th International Joint Conference on Natural Language Processing, EMNLP-IJCNLP 2019, Hong Kong, China, 3–7 November 2019, pp. 5349–5358 (2019). https://doi.org/10.18653/v1/D19-1538

Semantic-Aware Deep Neural Attention Network for Machine Translation Detection

Yangbin Shi[1,2], Jun Lu[2], Shuqin Gu[2], Qiang Wang[2], and Xiaolin Zheng[1(✉)]

[1] College of Computer Science and Technology, Zhejiang University,
Hangzhou, China
xlzheng@zju.edu.cn
[2] Machine Intelligence Technology Lab, Alibaba Group, Hangzhou, China

Abstract. Web crawling is an important way to collect a massive training corpus for building a high-quality machine translation system. However, a large amount of data collected comes from machine-translated texts rather than native speakers or professional translators, severely reducing the benefit of data scale. Traditional machine translation detection methods generally require human-crafted feature engineering and are difficult to distinguish the fine-grained semantic difference between real text and pseudo text from a modern neural machine translation system. To address this problem, we propose two semantic-aware models based on the deep neural network to automatically learn semantic features of text for monolingual scenarios and bilingual scenarios, respectively. Specifically, our models incorporate the global semantic from BERT and the local semantic from convolutional neural network together for monolingual detection and further explores the semantic consistency relationship for bilingual detection. The experimental results on the Chinese-English machine translation detection task show that our models achieve 83.12% F_1 in the monolingual detection and 85.53% F_1 in the bilingual detection respectively, which is better than the strong BERT baselines by 2.2–3.2%.

Keywords: Machine translation detection · Local & global semantic representation

1 Introduction

As we all know, data-driven machine translation, including statistical machine translation (SMT) [25] and neural machine translation (NMT) [4,23], strongly depends on the quality and quantity of the training corpora. For example, bilingual parallel pairs are used for supervision learning, and monolingual target data is available for language model[14] or data augmentation [20]. In practice, due

---

Supported in part by the National Key R&D Program of China (No. 2018YFB1403001).

J. Su and R. Sennrich (Eds.): CCMT 2021, CCIS 1464, pp. 63–76, 2021.
https://doi.org/10.1007/978-981-16-7512-6_6

to its low cost, data mining from subtitles and web crawling is one of the most popular ways to collect massive data for machine translation [12,19]. However, there are many noises in the collected data, which may mislead the model training and damage the performance of machine translation systems. In this work, we focus on the issue of *machine translation detection* (MTD) [1], which is a typical noise sourcing caused by the fact that a large amount of crawling data comes from machine-translated texts rather than native speakers or professional translators.

Most previous MTD work aims at SMT [1–3]. They design many human-crafted features and train binary statistical classifiers to identify whether a sentence comes from a SMT system. As SMT is notorious for long-distance reorder and is prone to generate the disfluent translation, these simple statistical classifiers can achieve good performance by adding some explicit linguistic features. However, the situation changes when turning to modern NMT systems. Specifically, NMT is modeled as a conditional language model, which is naturally good at generating fluent and grammatical translation [13]. Therefore, we argue that the previous coarse-grained MTD models cannot fit the NMT scenario, and it is necessary to design a fine-grained MTD model to distinguish the semantic bias between real text and machine-translated text.

To address this issue, we propose to model the deep semantic representation by neural network for both monolingual and bilingual scenarios. Specifically, aimed at monolingual sentence, we propose the **S**emantic-aware **I**nfluencing **A**ttention **N**etwork (SIAN) to capture the global and local semantic information of a sentence by combining BERT model[8] and Convolutional Neural Network (CNN) [11] together. SIAN integrates the important local semantic information into the global semantic information by adopting an influencing attention mechanism for obtaining the sufficient semantic representation of a sentence. In contrast, for the bilingual scenario, we further propose a **S**emantic **C**onsistency-aware **I**nteractive **A**ttention network (SCIA), which match the semantics of a target sentence with its corresponding source sentence to obtain the semantic consistency. In addition, the Part-of-Speech (POS) is used as the input to make the model better aware of the shallow syntactic information.

We compare our models with several baseline models (i.e., statistical classifier model and neural network-based models) on the outputs of four online NMT systems. Experimental results show that our proposed models outperform all of the baseline models by achieving an 83.12% F_1 in the monolingual scenario and an 85.53% F_1 in the bilingual scenario, respectively. To the best of our knowledge, we are the first to explore neural network-based techniques to tackle the machine translation detection task.

2   Related Work

Previous techniques for detecting machine-translated sentences are designed for SMT [1–3]. In the monolingual scenario, Arase et al. [3] designed a sentence-level classifier to distinguish the machine-translated sentences from a mixture of machine-translated and human-translated sentences. They utilized the phrase salad phenomenon and gappy-phrase features to detect if a sentence is fluent and grammatical. Aharoni et al. [1] utilized the common content-independent linguistic features for this detecting task. The features in their work were binary, denoting the presence or absence of each of a set of part-of-speech n-grams, as well as the function words. Both of their work adopted a binary statistical supervised classifier, i.e., SVM, to determine the likelihoods of machine-translated or human-translated sentences.

In the bilingual scenario, Antonova et al. [2] designed a phrase-based decoder for detecting machine-translated content in a Web-scraped parallel Russian-English corpus. By evaluating the BLEU score of translated content (by their decoder) against the target-side content, machine-translated content can be detected. Rarrick et al. [18] extract a variety of features, such as the number of tokens and character types, from sentences in both the source and target languages to capture words that are mistranslated by MT systems. With these features, the likelihood of a bilingual sentence pair being machine-translated can be determined.

The above work is designed for detecting the outputs of SMT by utilizing some explicit linguistic features and statistical supervised classifiers. We also address the problem as a binary classification task. In contrast, since the NMT has achieved significant success, we pay more attention to the implicit semantic features rather than such explicit linguistic features.

Data selection for machine translation system is a related area. These studies [5,7,14] aim to properly select data for training a subset sentence pairs from a large corpus, so that improve the performance of the MT system in the specific domain. Chen et al. [7] proposed a semi-supervised CNN based on bi-tokens (Bi-SSCNNs) for training machine translation systems from a large bilingual corpus. Moore et al. [14] use the language model to select domain-relate corpus. However, these methods are designed to select specific domain data. Our work utilizes the similar idea that detects the machine-translated sentences by relying on the neural networks for capturing more implicit information instead.

Another related area is the cross-lingual semantic textual similarity modeling, to which assesses the degree of two sentences in a different language is semantically equivalent to each other [6]. Shao et al. [21] use CNN to capture the semantic representation of the source and target sentences. Then a semantic difference vector between these two paired sentences is generated. While the aims of the tasks mentioned above are different from ours, we take the advantage of neural networks to obtain the semantic consistency information. We regard semantic consistency as an implicit feature for detecting the sentences with the semantic bias that were translated by the NMT system in the bilingual scenario.

3   Model Overview

This section,introduces our neural network-based methods for MTD task, including semantic-aware influencing attention network in monolingual scenario and semantic consistency-aware interactive attention network in bilingual scenario. The model architectures are shown in Fig. 1 and Fig. 2, respectively.

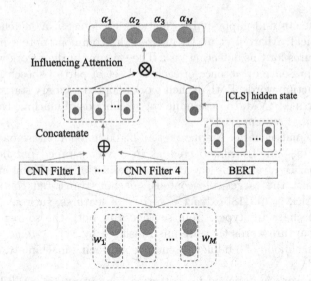

Fig. 1. Architecture of semantic-aware influencing attention network based on BERT and CNN (SIAN).

3.1 Semantic-Aware Influencing Attention Network (SIAN) in Monolingual Scenario

Our problem can be formulated as follows. Given a sentence with M words, we need to judge whether the sentence is machine-translated or human-translated. We propose a semantic-aware influencing attention network (SIAN) based on BERT and CNN for this task and the model architecture is shown in Fig. 1.

Global Semantic Feature Extraction by BERT. Specifically, the [CLS] token's hidden state is used as the hidden contextual representation of a sentence.

Local Semantic Feature Extraction by CNN. In order to capture the local semantic information of the sentence sufficiently, we use convolution blocks with different sizes of filters to encode the input sentence.

Let $w_i \in \mathbb{R}^d$ be the d-dimensional word vector corresponding to the i-th word in the sentence. Let $\mathbf{X} \in \mathbb{R}^{M \times d}$ denotes the input sentence where M is the length of the sentence with padding. A convolutional filter $\mathbf{W}_c \in \mathbb{R}^{d \times k}$ maps k words in the respective filed to a single feature c. As we slide the filter across the whole sentence, we obtain a sequence of new features $\mathbf{c} = [c_1, c_2, ..., c_M]$.

$$c_i = f(\mathbf{X}_{i:i+k} * \mathbf{W}_c + b_c), \tag{1}$$

where $b_c \in \mathbb{R}$ is a bias term and f is a nonlinear transformation function such as ReLU, $*$ denotes convolution operation.

SIAN Model for Machine Translation Detection. We have introduced the process about one feature is extracted from one filter. Since our SIAN model utilizes multiple filters with different filter sizes to generate multiple feature maps

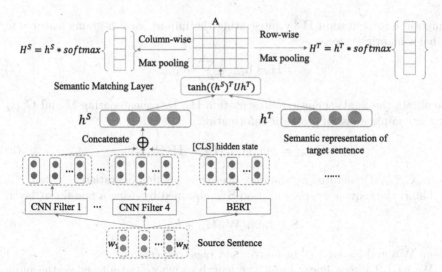

Fig. 2. Architecture of semantic consistency-aware interactive attention network (SCIA)

Table 1. An example of paired sentence.

Source	自由必须是有目标的自由，不然的话，我们便很容易感到厌倦
Human	Freedom must be **a purposeful freedom**, otherwise, we can easily get tired of it.
MT	Freedom must be *freedom of purpose*, otherwise we will easily get bored.

for capturing more local n-grams semantic information of a sentence. Therefore, we obtain the final local n-grams semantic representation by concatenating the different feature maps, $\mathbf{C} = [\mathbf{c}_1; \mathbf{c}_2; ...; \mathbf{c}_n]$, where n is the number of filters.

Moreover, we capture the global semantic representation by using the $[CLS]$ token's hidden state, $\mathbf{h}_{cls}$.

Next, we utilize the global semantic vector $\mathbf{h}_{cls}$ and the convolutional features vector $\mathbf{C}$ to calculate the attention weights, which attempts to capture some important local n-grams features to supplement the global semantic information:

$$\alpha_i = \frac{exp(s(\mathbf{c}_i, \mathbf{h}_{cls}))}{\sum_{j=1}^{M} exp(s(\mathbf{c}_j, \mathbf{h}_{cls}))} \quad (2)$$

where s is a score function that calculates the importance of $\mathbf{c}_i$ in the whole n-grams semantic features. The score function is defined as:

$$s(\mathbf{c}_i, \mathbf{h}_{cls}) = tanh(\mathbf{c}_i \cdot \mathbf{W}_a \cdot \mathbf{h}_{cls}^T + b_a) \quad (3)$$

where $\mathbf{W}_a$ and b_a are weight matrix and bias respectively, tanh is a non-linear function and $\mathbf{h}_{cls}^T$ is the transpose of the $\mathbf{h}_{cls}$. Then we can get the sufficient global semantic representation $\mathbf{H}$ by integrating the import local n-grams features to global semantic vector,

$$\mathbf{H} = \mathbf{h}_{cls} + \sum_{i=1}^{M} \alpha \mathbf{c}_i \tag{4}$$

We obtain the final semantic representation $\mathbf{H}_R$ by concatenating $\mathbf{H}$ and $\mathbf{C}_{max}$ for the completeness of semantic information,

$$\mathbf{C}_{max} = max(\mathbf{C}), \mathbf{H}_R = [\mathbf{H}; \mathbf{C}_{max}] \tag{5}$$

where $\mathbf{C}_{max}$ is generated by the max-over-time pooling operation.

Later, the sequence representation $\mathbf{S}$ is obtained by using a non-linear layer:

$$\mathbf{S} = tanh(\mathbf{W}_R \mathbf{H}_R + b_R), \tag{6}$$

where $\mathbf{W}_R$ and b_R are weight matrix and bias, respectively.

We feed $\mathbf{S}$ into a linear layer, the length of whose output equals the number of class labels. Finally, we add a *softmax* layer to calculate the probability distribution for judging a sentence is machine-translated or human-translated:

$$\mathbf{y} = softmax(\mathbf{W}_f \mathbf{S} + b_f), \tag{7}$$

where $\mathbf{W}_f$ and b_f are the weight matrix and bias of *softmax* layer, respectively.

3.2 Semantic Consistency-Aware Interactive Attention Network (SCIA) in Bilingual Scenario

We further tackle this detecting task from the perspective of semantic consistency in the bilingual scenario. For instance, given a source sentence, the standard human-translated sentence and machine-translated sentence of the target side are shown in Table 1.

In this example, due to the high performance of the NMT, we find that a machine-translated sentence is as fluent and grammatical as a sentence generated by human. When we focus on its semantics, we will find that its semantic information is a little different from its source sentence. Therefore, in order to better distinguish whether a sentence is machine-translated, we should further focus on whether its semantics is consistent with its corresponding source sentence.

Here, the BERT and CNN are also used to encode the global and local semantic representations in this scenario. We directly concatenate the representations of the BERT and CNN without pooling for source sentence S (similar to the target sentence T), generating the semantic vector $\mathbf{h}^S$ (semantic representation $\mathbf{h}^T$ for the target sentence). The architecture is shown in Fig. 2.

SCIA Model for Machine Translation Detection. In the bilingual scenario, we should pay more attention to the mutual semantic relation between the source and target sentence. Thus, an interactive attention network is proposed to capture semantic consistency.

Interactive attention is an approach that enables the semantic matching layer to be aware of the current input pair, in a way that the $\mathbf{h}^S$ is able to directly influence the $\mathbf{h}^T$, and vice versa. The main idea of the interactive attention is to encourage the hidden contextual representations interactively learning semantic matching information for the source and target sentences. Then the attention weights can be calculated by applying the column-wise and row-wise max pooling over $\mathbf{A}$ matrix.

Consider the input pair (S, T) where the length of the source sentence S is N, and the length of the target sentence T is M. The matrix $\mathbf{A} \in \mathbb{R}^{N \times M}$ can be calculated as follows:

$$\mathbf{A} = tanh((\mathbf{h}^S)^T \mathbf{U} \mathbf{h}^T + b_A), \tag{8}$$

where $\mathbf{U}$ is a weight matrix, b_A is the bias, and $(\mathbf{h}^S)^T$ denotes the transpose of the $\mathbf{h}^S$.

Later, we apply the column-wise and row-wise max pooling over the $\mathbf{A}$ matrix to generate the vectors $\mathbf{a}^s \in \mathbb{R}^N$ and $\mathbf{a}^t \in \mathbb{R}^M$, respectively.

$$[a^s]_i = \max_{1 < n < N}[\mathbf{A}_{i,n}] \tag{9}$$

$$[a^t]_i = \max_{1 < m < M}[\mathbf{A}_{m,i}] \tag{10}$$

Each element i of the vector $\mathbf{a}^t$ can be interpreted as an importance for the local n-grams semantic information around the i-th word in the representation of target sentence $\mathbf{h}^T$ according to the representation of source sentence $\mathbf{h}^S$. In the same way, each element i of the vector $\mathbf{a}^s$ can be interpreted as an importance for the local n-grams semantic information around the i-th word in the representation of source sentence $\mathbf{h}^S$ according to the representation of the target sentence $\mathbf{h}^T$.

Sequentially, we adopt the *softmax* function to the vectors $\mathbf{a}^s$ and $\mathbf{a}^t$ to generate the attention weight α and β

$$[\alpha^s]_i = \frac{exp([a^s]_i)}{\sum_{1 < b < M} exp([a^s]_b)} \tag{11}$$

$$[\beta^t]_i = \frac{exp([a^t]_i)}{\sum_{1 < b < N} exp([a^t]_b)} \tag{12}$$

Next, we can get the final representations of the source and target sentences, respectively:

$$\mathbf{H}^S = \mathbf{h}^S * \alpha \tag{13}$$

$$\mathbf{H}^T = \mathbf{h}^T * \beta \tag{14}$$

In addition, we apply element-wise absolute difference and element-wise dot product, which model the semantic bias information and consistency information between two semantic vectors ($\mathbf{H}^S$ and $\mathbf{H}^T$), respectively.

$$\mathbf{H}_i^{(1)} = |\mathbf{H}_i^S - \mathbf{H}_i^T| \tag{15}$$

Table 2. Chinese-English data-sets

ZH-EN	Train-set	Development-set	Test-set
Human-trans	$3.7 * 10^6$	5000	5000
Machine-trans	$3.7 * 10^6$	5000	5000

$$\mathbf{H}_i^{(2)} = \mathbf{H}_i^S \odot \mathbf{H}_i^T \tag{16}$$

The final semantic consistent vector is got by concatenating $\mathbf{H}^{(1)}$ and $\mathbf{H}^{(2)}$, $\mathbf{H}_F = [\mathbf{H}^{(1)}; \mathbf{H}^{(2)}]$, and then feed it to a fully connected layer to get the probability distribution for judging a sentence is machine-translated or human-translated.

4 Experiments

4.1 Data Preparation

For the purpose of evaluation, we use human-translated and machine-translated sentences to train our proposed models. For the human-translated sentences, we use the WMT18 Chinese-English (*ZH-EN*) parallel sentence pairs[1]. A few methods [9] are used to filter the *lower-quality* sentence pairs. For the machine-translated sentences, we randomly feed the source sentences (i.e., the Chinese sentences) of the above high-quality parallel corpus to four online commercial machine translators[2] for obtaining target sentences (i.e., English sentence). In this way, we can obtain large amounts of positive and negative (i.e., human- and machine-translated sentence pairs) data instances. Moreover, the source sentences are segmented and POS tagged by using an NLP toolset we developed. The target sentences are tokenized and POS tagged by using NLTK toolset[3]. The whole data-set is divided into three parts: train-set, development-set and test-set. Table 2 shows the details of the data-sets used in our experiments.

4.2 Model Parameters Settings

In our experiments, including monolingual and bilingual scenarios, the POS tags are converted to the corresponding tag embeddings, and the dimension of the word embedding and the POS tag embedding are both set to 300, all of them are randomly initialized and updated during the training process. In addition, four convolution blocks are used with kernel windows of $1, 2, 3, 4$, each with 200 feature maps. And in order to have a similar number of parameters in the CNN, the BERT model is set to be 512 hidden size and 12 layers. We use PyTorch[4] to implement our proposed models and employ the Adadelta [24] as the training algorithm, whose decay rate is set to 0.95. The regularization parameter λ is set to 10^{-4} and the initial learning rate is set to 1.0.

---

[1] http://www.statmt.org/wmt18/translation-task.html.
[2] To our knowledge, all the four machine translators are NMT systems.
[3] https://www.nltk.org/.
[4] https://pytorch.org/.

Table 3. Performance of models in the monolingual scenario.

Model	Acc	F_1
SVM	70.93	70.84
CNN	73.31	73.79
BERT	80.01	79.89
CNN+POS	74.78	74.31
SN	81.76	81.54
SIAN	82.45	82.56
SIAN+POS	**83.01**	**83.12**

4.3 Evaluation Metric

To evaluate our models, we adopt the *Accuracy* (*Acc*) and F_1 score as metrics, where $Acc = \frac{number\ of\ correct\ predictions}{Total\ number\ of\ predictions}$ and $F_1 = \frac{2*precision*Recall}{precision+Recall}$.

4.4 Model Comparison and Analysis in Monolingual Scenario

In order to evaluate the performance of our SIAN model, we compare it with the statistic classifier, i.e., SVM, and the CNN/BERT models used in data selection.

SVM: Using the common content-independent linguistic features, such as N-grams, function words and POS tags, and adopt the SVM-SMO as a classifier for this detecting task [1].

CNN/BERT: Using CNN or BERT as a sentence encoder, and then stack two fully connected layers.

SN: Semantic-aware network (SN) model is designed by us in this work, which also adopts CNN and BERT to encode the local and global semantic information of a sentence. The only difference between SN and SIAN is that SN does not utilize the influencing attention mechanism.

Table 3 shows the performance of our SIAN model and other methods. It is obvious that our SIAN model with POS tags achieves the best performance among all methods. We can find that SVM gets the worst performance because this method mainly depends on some linguistic features or rules to judge the fluency degree of a sentence for detecting the outputs of the SMT. The quality of translations in NMT has been improved significantly over SMT, and the fluency of NMT generated sentences are close to the human-translated sentences. Thus machine-translated sentences cannot be effectively identified if only rely on such features and classifiers.

Furthermore, when we compare the SN model with CNN and BERT models, we find that SN model achieves better performance. Because if only adopt CNN or BERT as a sentence encoder, which may neglect the global semantics or local semantics of a sentence, while SN model simultaneously takes into account this

Table 4. Performance of models in the bilingual scenario.

Model	Acc	F_1
CNN	78.59	76.32
BERT	83.24	83.31
CNN-Pair	80.19	78.90
CNN-Pair+POS	80.62	79.36
SCN	84.12	84.32
SCIA	84.98	84.87
SCIA+POS	**85.35**	**85.53**

semantic information instead. Therefore, according to these three experimental results, we can demonstrate that it is important to combine the local and global semantic features in this task.

As for the SIAN model, it outperforms the SN model. Since it pays more attention to some important local n-grams semantic information that is achieved by the influencing attention mechanism. Besides, SIAN integrates the local semantic information into the global semantic information to obtain the sufficient semantic representation of a sentence.

Here, we further employ the shallow syntactic information (i.e., POS) of the sentences as an auxiliary feature to improve the performance of this task. From Table 3, we can find that models with the POS tags perform better than their corresponding models without POS tags.

4.5   Model Comparison and Analysis in the Bilingual Scenario

In this subsection, we compare the SCIA model with the following models.

CNN/BERT: CNN or BERT are used to encode source and target sentences. Then the encoding vectors of the sentence pair are concatenated to generate the final representation. Finally, we apply two fully connected layers to compute a unique score for a bilingual sentence pair [16].

CNN-Pair: Using CNN to capture the semantic vectors of the source and target sentences, respectively. Then generates a semantic difference vector between a sentence pair by concatenating their element-wise absolute difference and the element-wise multiplication of their semantic vectors. Finally, the feed-forward layer is used to obtain a similarity score [21].

SCN: Semantic consistency-aware network (SCN) model is designed by us in this scenario, whose architecture is similar to the SCIA model. The only difference between these two models is that the SCN model does not utilize the interactive attention mechanism.

From Table 4, it is obvious that our SCIA model with POS tags achieves the best performance. We can find that the CNN-Pair model performs better than

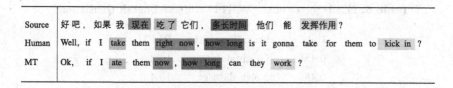

Fig. 3. A real case from our test set.

CNN model because both CNN encode the representations of the source and target sentences without considering the semantic bias of the paired sentences. Instead, the CNN-Pair model takes advantage of the element-wise absolute difference and the element-wise multiplication of the corresponding paired sentence level embedding. It can model the relation of the source and target sentence and is conducive to identify machine-translated sentences. Although CNN-Pair model considers the relationship between the sentence pairs, it only captures the local semantic information of the source and target sentences while without taking the global semantic information into account. Thus CNN-Pair model performs less competitively than our SCN model.

As for the SCIA model, it outperforms the SCN model since the SCIA realizes the importance of the mutual relationship between a source and target sentence pair by utilizing the interactive attention mechanism. It enables the semantic matching layer to be aware of the current input pair in a way that the current semantic representation of the source sentence can directly influence the semantic representation of the target sentence and vice versa. Thus, the SCIA model can learn more semantic consistency information than the SCN model. Similar to the monolingual scenario, the POS tags bring further improvement to the CNN-Pair or SCIA model.

4.6 Case Study

Particularly, to have an intuitive understanding of our proposed model, we give a sample instance to illustrate the characteristics of the SCIA model better as shown in Fig. 3. The same color corresponds to the word alignment translation. From this case, we can find that although the machine-translated sentence can be translated accurately in the word alignment level, its semantics of the whole sentence is ambiguous according to its corresponding source sentence, i.e., there is some semantic bias compared with the corresponding source sentence. Thus, the statistical classifiers tend to identify these sentences as human-translated while the SCIA model does not.

4.7 Evaluation on Neural Machine Translation Systems

We further test our SCIA model on an NMT system [4,22].

The experiments are carried out with an open-source system called Marian [10], which is a transformer-based NMT training system[23]. We carry out exper-

Table 5. BLEU scores of the WMT17 Chinese-English translation.

Data size	Data description	BLEU
0.4M	Original dataset	16.3
0.4M	Noisy dataset	15.4
0.34M	Clean-up dataset	15.9

iments on the Chinese-English dataset of WMT2017 task[5]. We select 400000 sentence pairs from these datasets as the original training dataset; the development set is WMT2017's test set, which contains 2002 sentence pairs. The test set comes from WMT2018 news translation task, which contains 3981 sentence pairs. Then we randomly select 30% sentences from the training data and obtain the corresponding machine-translated target sentences by four online machine translators, obtaining noisy dataset. Next, we use our proposed SCIA model to filter out the machine-translated sentence pairs from the noisy dataset, obtaining the clean-up dataset.

Table 5 shows the BLEU[15] scores of the NMT systems based on different training data. From this table, we can see that when we introduce the noise to the original data, we lost 0.9 BLEU score. Then, if we apply our SCIA model to the noisy data, the BLEU score improves the performance to 15.9 on the clean-up data, which demonstrates that the SCIA model can screen out the machine-translated sentences for improving the performance of the NMT system.

The Back-Translation method [17, 20] has been widely used in building NMT systems. Our models may improve the performance of Back-Translation further by filtering low-quality back translated sentence pairs.

5   Conclusion

In this paper, we propose two neural network models for detecting the sentences generated by NMT in monolingual and bilingual scenarios, including a semantic-aware influencing attention network (SIAN), which is used to capture important local semantic information; and a semantic consistency-aware interactive attention network (SCIA), which is used to capture semantic matching between a source and target sentence pair. Results show that our models outperform all of the baseline models by achieving an 83.12% F_1 in the monolingual scenario and an 85.53% F_1 in the bilingual scenario respectively, which is better than the strong BERT baselines by 2.2–3.2%. To the best of our knowledge, SIAN and SCIA are the first neural network-based models that are proposed to apply on the NMT output detection task.

---

[5] http://www.statmt.org/wmt17/translation-task.html.

References

1. Aharoni, R., Koppel, M., Goldberg, Y.: Automatic detection of machine translated text and translation quality estimation. In: Proceedings of the 52nd Annual Meeting of the Association for Computational Linguistics (Volume 2: Short Papers), vol. 2, pp. 289–295 (2014)
2. Antonova, A., Misyurev, A.: Building a web-based parallel corpus and filtering out machine-translated text. In: Proceedings of the 4th Workshop on Building and Using Comparable Corpora: Comparable Corpora and the Web, pp. 136–144. Association for Computational Linguistics (2011)
3. Arase, Y., Zhou, M.: Machine translation detection from monolingual web-text. In: Proceedings of the 51st Annual Meeting of the Association for Computational Linguistics (Volume 1: Long Papers), vol. 1, pp. 1597–1607 (2013)
4. Bahdanau, D., Cho, K., Bengio, Y.: Neural machine translation by jointly learning to align and translate. arXiv preprint arXiv:1409.0473 (2014)
5. Biçici, E., Yuret, D.: Instance selection for machine translation using feature decay algorithms. In: Proceedings of the Sixth Workshop on Statistical Machine Translation, pp. 272–283. Association for Computational Linguistics (2011)
6. Cer, D., Diab, M., Agirre, E., Lopez-Gazpio, I., Specia, L.: SemEval-2017 task 1: semantic textual similarity-multilingual and cross-lingual focused evaluation. arXiv preprint arXiv:1708.00055 (2017)
7. Chen, B., Kuhn, R., Foster, G., Cherry, C., Huang, F.: Bilingual methods for adaptive training data selection for machine translation. In: Proceedings of AMTA, pp. 93–103 (2016)
8. Devlin, J., Chang, M.W., Lee, K., Toutanova, K.: BERT: pre-training of deep bidirectional transformers for language understanding. arXiv preprint arXiv:1810.04805 (2018)
9. Eetemadi, S., Lewis, W., Toutanova, K., Radha, H.: Survey of data-selection methods in statistical machine translation. Mach. Transl. **29**, 189–223 (2015). https://doi.org/10.1007/s10590-015-9176-1
10. Junczys-Dowmunt, M., et al.: Marian: fast neural machine translation in C++. In: Proceedings of ACL 2018, System Demonstrations, pp. 116–121. Association for Computational Linguistics, Melbourne, July 2018. http://www.aclweb.org/anthology/P18-4020
11. Kim, Y.: Convolutional neural networks for sentence classification. arXiv preprint arXiv:1408.5882 (2014)
12. Lison, P., Tiedemann, J.: OpenSubtitles 2016: extracting large parallel corpora from movie and tv subtitles (2016)
13. Ma, M., Nirschl, M., Biadsy, F., Kumar, S.: Approaches for neural-network language model adaptation. In: Proceedings of Interspeech, Stockholm, Sweden, pp. 259–263 (2017)
14. Moore, R.C., Lewis, W.: Intelligent selection of language model training data (2010)
15. Papineni, K., Roukos, S., Ward, T., Zhu, W.J.: BLEU: a method for automatic evaluation of machine translation. In: Proceedings of the ACL, pp. 311–318 (2002). https://doi.org/10.3115/1073083.1073135. http://dx.doi.org/10.3115/1073083.1073135
16. Peris, Á., Chinea-Ríos, M., Casacuberta, F.: Neural networks classifier for data selection in statistical machine translation. Prague Bull. Math. Linguist. **108**(1), 283–294 (2017)

17. Poncelas, A., Shterionov, D., Way, A., Wenniger, G.M.D.B., Passban, P.: Investigating backtranslation in neural machine translation. arXiv preprint arXiv:1804.06189 (2018)
18. Rarrick, S., Quirk, C., Lewis, W.: MT detection in web-scraped parallel corpora. In: Proceedings of the Machine Translation Summit (MT Summit XIII) (2011)
19. Resnik, P., Smith, N.A.: The web as a parallel corpus. Comput. Linguist. **29**(3), 349–380 (2003)
20. Sennrich, R., Haddow, B., Birch, A.: Improving neural machine translation models with monolingual data. arXiv preprint arXiv:1511.06709 (2015)
21. Shao, Y.: HCTI at SemEval-2017 task 1: use convolutional neural network to evaluate semantic textual similarity. In: Proceedings of the 11th International Workshop on Semantic Evaluation (SemEval-2017), pp. 130–133 (2017)
22. Sutskever, I., Vinyals, O., Le, Q.V.: Sequence to sequence learning with neural networks. In: Advances in Neural Information Processing Systems, pp. 3104–3112 (2014)
23. Vaswani, A., et al.: Attention is all you need. In: Advances in Neural Information Processing Systems, pp. 5998–6008 (2017)
24. Zeiler, M.D.: ADADELTA: an adaptive learning rate method. arXiv preprint arXiv:1212.5701 (2012)
25. Zens, R., Och, F.J., Ney, H.: Phrase-based statistical machine translation. In: Jarke, M., Lakemeyer, G., Koehler, J. (eds.) KI 2002. LNCS (LNAI), vol. 2479, pp. 18–32. Springer, Heidelberg (2002). https://doi.org/10.1007/3-540-45751-8_2

Routing Based Context Selection for Document-Level Neural Machine Translation

Weilun Fei[1], Ping Jian[1,2(✉)], Xiaoguang Zhu[1], and Yi Lin[1]

[1] School of Computer Science and Technology, Beijing Institute of Technology, Beijing 100081, China
{wlfei,pjian,xgzhu,1120180804}@bit.edu.cn
[2] Beijing Engineering Research Center of High Volume Language Information Processing and Cloud Computing Applications, Beijing Institute of Technology, Beijing 100081, China

Abstract. Most of the existing methods of document-level neural machine translation (NMT) integrate more textual information by extending the scope of sentence encoding. Usually, the sentence-level representation is incorporated (via attention or gate mechanism) in these methods, which makes them straightforward but rough, and it is difficult to distinguish useful contextual information from noises. Furthermore, the longer the encoding length is, the more difficult it is for the model to grasp the inter-dependency between sentences. In this paper, a document-level NMT method based on a routing algorithm is presented, which can automatically select context information. The routing mechanism endows the current source sentence with the ability to decide which words can become its context. This leads the method to merge the inter-sentence dependencies in a more flexible and elegant way, and model local structure information more effectively. At the same time, this structured information selection mechanism will also alleviate the possible problems caused by long-distance encoding. Experimental results show that our method is 2.91 BLEU higher than the Transformer model on the public dataset of ZH-EN, and is superior to most of the state-of-the-art document-level NMT models.

Keywords: Natural Language Processing · Document-Level Neural Machine Translation · Routing Algorithm

1 Introduction

With the development of deep learning methods, neural machine translation (NMT) has made remarkable progress in most language pairs. However, the standard NMT methods are first designed for sentence-level [1–3], which may bring some document-level errors, such as document inconsistency [4–9]. In order to reduce the errors caused by sentence-level NMT when translating discourses,

© Springer Nature Singapore Pte Ltd. 2021
J. Su and R. Sennrich (Eds.): CCMT 2021, CCIS 1464, pp. 77–91, 2021.
https://doi.org/10.1007/978-981-16-7512-6_7

a large number of document-level NMT methods have been proposed to improve the translation performance by using context outside a single sentence.

The most recent context-aware methods take the context of the current sentence as inputs of NMT model, and attach another input stream in parallel [4,5,10]. Therefore, most researchers tend to reform the mature NMT models to merge the representation from previous sentences as context [4,5,11–14] into every layer of the encoder or decoder to consider the information from cross sentences. To improve the comprehension of the current text, people can combine with the future context. It is very common for us, not to mention the neural network lacking prior knowledge and common sense. Consequently, context is not necessarily limited to the sentences before the current sentence, it can also come from the future, which is ignored but effective. However, if we simply and roughly expand the scope of sentences as inputs without filtering them, it may bring burden to the model. According to [15], information in the context is not always useful. We are supposed to increase the content of context selectively.

This paper draws lessons from a routing method [16] of multilingual NMT (MNMT), and puts forward a document-level NMT routing method based on this algorithm. In MNMT, researchers find that using a mix of shared and language-specific parameters can help the models obtain a great improvement in exploring universal MNMT, but keep the question of when and where language-specific capacity matters most. This is similar to what kind of context is the most useful in document-level NMT. According to [15], we can assume that every word in the context contains different levels of document-aware information. In order to filter redundant information of context, we use routing algorithm, which helps the model select words whose document-level information is more important as context automatically. On the one hand, we avoid long-distance encoding. On the other hand, redundant contextual information is filtered out.

In our experiments, we choose the sentence before the current sentence and the sentence next to the current sentence as context. The results show that the changes we made improve the performance of document-level NMT. Compared with the methods which utilize the whole document as context [9,17], our method still has competitiveness, especially on the dataset of ZH-EN.

2   Related Work

With the latest development and performance improvements of neural networks, people are more interested in document-level MT and textual context also shows its importance to machine translation. Based on the encoder-decoder NMT framework, existing works mainly use the following three methods to introduce document-level information:

Single-Encoder Approach. This kind of method expands the range of sentences when inputting them into the model, such as [6,18–20], which has done a lot of research about the input of model, including the expansion of encoder input and decoder input. This kind of method is relatively rough for the application of context, which is the earliest attempt of encoder-decoder framework. These attempts proved that not only the previous context but also the future context can improve the translation effect, which is gradually ignored in later studies. In addition, the method of fusing context at the encoder side contributes more than the method of fusing context at the decoder side. Because fusion at the decoder side may lead to error propagation.

Multi-Encoder Approach. According to when and where to fuse the output of the multi-encoder inside the decoder (see Fig. 1), [13] or outside the decoder (see Fig. 2), [13,21,22]. Reference [23] divides the multi-encoder method into inside multi-encoders [4,19,24,25] and outside multi-encoders [5,9,17,26]. The Multi-encoder method mainly adopts two fusion methods: 1) Some methods use attention mechanisms to encode context statements into the encoder or decoder, for example, Reference [4] inserts a context-attention layer into the model; 2) the others use the gate mechanism to aggregate context, thus learning anaphora resolution. These methods are similar in that they all add Transformer models with additional context-related modules.

Post Processing. Reference [27] uses deliberation network, which adds another decoder after Transformer, and employs reward teacher to model coherence for document-level machine translation. Reference [8] uses another method called document-level repair, which makes full use of monolingual document-level data in the target language.

Inspired by previous works, we add an extra context module to the Transformer model to extract context information. Reference [15] suggests that in document-level NMT, sometimes context is too long to simplify calculations, and in fact, a lot of information in the context is actually unnecessary. They retain the most likely words of the context, such as named entities and special words like POS. Combined with the above points, we use the routing method in multilingual NMT, and hope that the model itself can combine the input sentence to determine which words are useful for forming context, not just named entities and POS, so as to improve the translation effect.

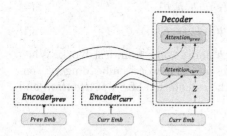 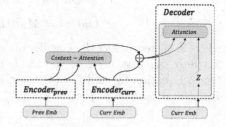

Fig. 1. Fusion inside the decoder   **Fig. 2.** Fusion outside the decoder

3   Background

3.1   Document-Level NMT

Compared with sentence-level NMT, document-level NMT considers contextual information. We assume that $(X, Y) \in \mathbb{C}$ where X represents source sentences and Y represents target sentences. We use x^i to express the i-th sentence in X. y^i denotes the i-th sentence in Y. In order to generate target sentence y^k, document-level NMT is supposed to make full use of the contextual information of source sentence x^k. As the input of encoder, x is converted into the hidden state H. We define set $X^{<>k}$ as context of x^k, and then we can approximate the document-level translation probability as:

$$P(y^k|x^k;\theta) = \prod_{i=1}^{n} p(y_i^k|y_{<i}, H^k, X^{<>k};\theta) \tag{1}$$

3.2   Transformer

Encoder-Decoder architecture composed of sequence models, like RNN or LSTM, has made great improvement in NMT [2,3]. However, Transformer [28], which relies entirely on attention mechanism, has surpassed most previous models. Considering the above points, we choose Transformer as our basic model.

To avoid gradient vanishing or explosion, the following residual normalization structure is used for the Transformer block:

$$z = LayerNorm(h + f(h)) \tag{2}$$

where h represents the output from the last block, z is the output of this block, $LayerNorm(\cdot)$ means Layer Normalization and $f(\cdot)$ can be MultiHead Attention or Feed-Forward Network. The encoder of Transformer includes Multi-Head Self-Attention and Feed-Forward Network. Though the decoder has similar sub-layers, another sub-layer called Encoder-Decoder Attention is inserted between these layers. With the help of MultiHead Attention, the model can pay attention to information from different representation subspaces:

$$Output = MultiHead(h, h, h) \tag{3}$$

$$Output = MultiHead(z, E_{out}, E_{out}) \tag{4}$$

where E_{out} represents the encoder output, h and z come from last block. When fed into the first layer of model, h represents the word embedding of sentence. Equation (3) and Eq. (4) stand for the calculation of *Self Attention* and *Enc Dec Attention* respectively.

3.3   Conditional Language-Specific Routing (CLSR)

From the perspective of the mapping between language pairs, the MNMT model has three strategies: many-to-one, one-to-many and many-to-many. Reference [29] raises the question that just using specific language signs is not enough to explore the features of specific language. To make a thorough inquiry of when and where language specific modeling matters most in MNMT, reference [16] introduces conditional language-specific routing (CLSR), a method that keeps the balance between language-specific path and shared path as controlled by the gates. Equation (2) can be modified as follows:

$$z = LayerNorm(h + CLSR(f(h)))\qquad(5)$$

CLSR learns a gate $g(\cdot)$ for each input token, which helps blocks in Transformer selectively route information through language-specific path h^{lang} or shared path h^{shared}:

$$CLSR(f(h)) = g(h) \odot h^{lang} + (1 - g(h)) \odot h^{shared}\qquad(6)$$

$$h^{lang} = f(h)W^{lang},\ \ h^{shared} = f(h)W^{shared}\qquad(7)$$

where W^{shared} represents the trainable parameters shared across languages and W^{lang} is the trainable parameters for specific languages. The gate $g(\cdot)$ is computed from a two-layer feed-forward network $G(\cdot)$, and zero-mean Gaussian noise is used to discretize it during training:

$$g(h) = \sigma(G(h) + \alpha(t)\mathcal{N}(0, 1))\qquad(8)$$

$$G(h) = Relu(hW_1 + b)W_2\qquad(9)$$

where $\sigma(\cdot)$ is the logistic-sigmoid function, and W_1 as well as W_2 is trainable parameters. $\alpha(\cdot)$ is a linearly function and increases with training step t. When inferencing, $g(h)$ is replaced with a decision rule: $g(h) = \delta(G(h) > 0)$, where $\delta(\cdot)$ is a Dirac measure.

4   Method

In this section, we will introduce how we apply the aforementioned routing algorithm to selecting words as context automatically for document-level NMT in detail. Before that, we will introduce the symbols used in the model.

Assuming X and Y represent the source and target sentences in corpus $\mathbb{C}$. We define that $c_{k_{-1}}^l, c_{k_1}^l$ are outputs from the l-th Prev Encoder Layer and Post Encoder Layer. $x_{k_{-1}}^l, x_{k_1}^l$ and x_k^l are the input of the l-th encoder layer. When $l = 0$, $x_k^l = x_k$, it's the same as $x_{k_{-1}}^l$ and $x_{k_1}^l$. c_k^l means the l-th layer context hidden state, which is got by gate aggregation. $x_{k,self-attn}^l$ is used to represent the output from self-attention layer of l-th layer. We can see the details of the model in Fig. 3.

4.1   Inputs of Our Model

Considering the differences between sentence-level NMT and document-level NMT, it's necessary to introduce the inputs composition of our model. In Transformer, researchers use sine and cosine functions to calculate position embedding, which helps the attention mechanism pay attention to the word position information added to the word embedding. While in document-level NMT, the

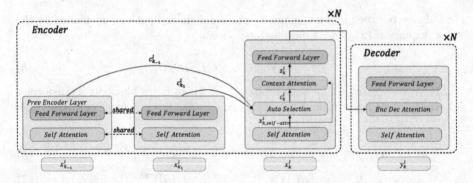

Fig. 3. The main architecture of our model. The two context-encoders share parameters. *Auto-Selection* Layer takes $x^l_{k,self-attn}$, c^l_{k-1} and $c^l_{k_1}$ as input to compute c^l_k which represents context information. To help $x^l_{k,self-attn}$ attend over all positions in the input context, Context Attention Layer takes $x^l_{k,self-attn}$ as query and c^l_k as key and value, in which case, we can get z^l_k.

information of sentences order has its significance. We refer to the idea of [10], in which way, we add the segment embedding to the position embedding and the word embedding. In Fig. 4, we take x^0_k as an example, which is the input of the first layer of our model, and we add different segment embedding to x^0_{k-1} and $x^0_{k_1}$ (0 indicates the previous context, 2 indicates the future context).

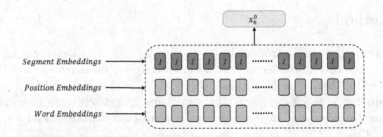

Fig. 4. The details about the composition of the inputs of the first layer of encoders, taking x^0_k as example, segment embeddings are set as 1 to represent the current sentence.

4.2   Context Attention

First of all, we explain how the model integrates context into the translation sentence. Between self-attention layer and feed-forward layer of the encoder for x_k, we insert *context attention* layer, which is defined as follows:

$$h_{AS} = MultiHead(x^l_{k,self-attn}, \ c^l_k, \ c^l_k) \tag{10}$$

$$z_k^l = LayerNorm(x_{k,self-attn}^l + h_{AS}) \tag{11}$$

where z_k^l is the output of this layer and context hidden state c_k^l is computed by the algorithm following, which will be introduced in detail.

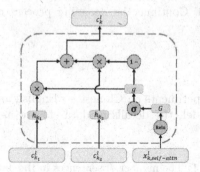

Fig. 5. The detail of *Auto-Selection* Layer. Gate is computed by $x_{k,self-attn}^l$

4.3   Auto-selection

Since we choose the sentences around x as context, we must filter out the information that may bring unnecessary noise from context. Inspired by CLSR, we hope that our model can help x keep the balance between the information from different sentences and filter out noise, just as CLSR helps MNMT models to decide when and where to use language-specific parameters or shared parameters:

$$c_k^l = g(x_{k,self-attn}^l) \odot h_{k-1} + (1 - g(x_{k,self-attn}^l)) \odot h_{k_1} \tag{12}$$

$$with \quad h_{k-1} = c_{k-1}^l W_{k-1}, \ h_{k_1} = c_{k_1}^l W_{k_1} \tag{13}$$

W_{k-1} as well as W_{k_1} is trainable parameters.

$$G(x_{k,self-attn}^l) = Relu(x_{k,self-attn}^l W_{-1} + b)W_1 \tag{14}$$

$$g(x_{k,self-attn}^l) = \sigma(G(x_{k,self-attn}^l)) \tag{15}$$

we can see Fig. 5 for details.

Compared with CLSR, we git rid of the zero-mean Gaussian noise, and totally let $x_{k,self-attn}^l$ itself to design its context. About other aspects of computation for $g(\cdot)$, following the configuration of CLSR, we apply a two-layers feed-forward network and use $Relu(\cdot)$ and $\sigma(\cdot)$ as activation function.

Now, we can summarize our training process as follows:

– The inputs $x_{k-1}^l, x_{k_1}^l$ are fed into the Prev/Post Encoder Layer respectively. The Prev Encoder Layer shares parameters with the Post Encoder Layer.

- The outputs of the Prev/Post Encoder Layer are sent to the encoder of x_k^l and x_k^l is used to calculate gate, which helps the model integrate more useful information from context to calculate the output of the encoder.
- The steps above are repeated for N times, i.e. the number of layers. Then, the outputs of encoder are sent to the decoder.
- Training the model. Continue the decoding process until meeting the end token.

5   Experiments

We mainly conduct experiments on Chinese $\rightarrow$ English and English $\rightarrow$ German task to verify our model, the details of datasets are as follows and listed in Table 1.

Table 1. The number of sentences in the datasets

Datasets		Training	Dev	Test
ZH-EN	TED	0.20M	0.88K	5.47K
EN-DE	TED	0.20M	8.96K	2.26K
	NEWS	0.22M	2.16K	2.99K
	Europarl	1.66M	3.58K	5.13K

5.1   Datasets

For fair comparison, we choose four widely used document-level parallel datasets, one Chinese $\rightarrow$ English dataset and three English $\rightarrow$ German datasets:

- TED (ZH-EN, TED). The Chinese $\rightarrow$ English datasets are from IWSLT 2015, where we mainly conduct our experiments. Following the work of [9], we take dev2010 as development set and tst2010-2013 as test set.
- TED (EN-DE, TED). According to [21], we choose IWSLT17 [30] as datasets for training. Tst2016-2017 is test set and the rest is the development set.
- News-Commentary (EN-DE, NEWS). Following [9] and [21], we obtain News Commentary v11 for training, WMT newstest 2015 for developing and WMT newstest2016 for testing.
- Europarl (EN-DE, Europarl). Train set, development set and test set are extracted from the Europarl v7 [31]. Details are mentioned in [21].

For TED ZH-EN dataset, we first use jieba for word segmentation. In all translation tasks, we tokenize the data with MOSE tokenizer [32] and apply byte-pair-encoding (BPE) algorithm [33] to encode words with sub-word units. We also use tools offered by Fairseq [34] to preprocess all dataset, in which form that model can accept.

5.2 Training Detail

On the basis of source code provided by Fairseq [34], we show detailed strategies for training the model. Adam [35] is the optimizer of the network with ($\beta_1 = 0.9, \beta_2 = 0.98$). $Warmup - updates$ is set 8000, $dropout$ is 0.1, where $warmup - init - lr$ is 10^{-7}. We set the batch size to 25,000 per batch and limit sentence length to 150 BPE tokens. For models on TED Zh-En, hidden dimension is $d_z = 256$, and the feed-forward dimension is $d_{ffn} = 512$. We use 4 layers in the encoder and decoder, each layer has 8 heads of attention. For the reset datasets, the hidden dimension and feed-forward dimension are set to 512/2048 respectively. Note that the above hyper-parameter settings are the same as those used in the baseline models.

5.3 Main Results

To make the results fair, we follow the work of [9] and [21] who use sacrebleu [36] to evaluate the translation quality. In addition to the baseline Transformer, we also compare our model with five state-of-the-art document-level NMT models including:

- Document-aware Transformer (DocT, [4]). Introducing context information by adding context sub-layers at each encoder and decoder layer.
- Hierarchical Attention NMT (HAN, [13]). Capturing the context in a structured and dynamic manner.
- Selective Attention NMT (SAN, [21]) Using sparse attention to selectively focus on relevant sentences.
- Query-guided Capsule Network (QCN, [22]). Clustering context information into different perspectives from which the target translation may concern.
- Arbitrary Context NMT (ACN, [9]). Being able to deal with documents containing any number of sentences.

Table 2. BLEU results on four datasets. The score in parentheses represents the BLEU of their baseline.

#	Models	ZH-EN	EN-DE		
		TED (baseline)	TED (baseline)	NEWS (baseline)	Europarl (baseline)
1	DocT (2018) [4]	n/a	24.00 (23.28)	23.08 (22.78)	29.32 (28.72)
2	HAN (2018) [13]	17.90 (17.00)	24.58 (23.28)	25.03 (22.78)	28.60(28.72)
3	SAN (2019) [21]	n/a	24.42 (23.28)	24.84 (22.78)	29.75 (28.72)
4	QCN (2019) [22]	n/a	25.19 (23.28)	22.37 (21.67)	29.82 (28.72)
5	ACN (2020) [9]	19.10 (17.00)	25.10 (23.10)	24.91 (22.40)	30.40 (29.40)
Ours					
6	Transformer (2017) [28]	17.11	23.20	23.13	29.49
7	**Our model**	20.02	25.01	24.03	29.87

As shown in Table 2, the proposed model improves the BLEU scores of the aforementioned datasets by 2.91, 1.81, 0.90 and 0.38 points compared with the baseline of sentence-level Transformer. Especially on TED ZH-EN, our model makes a significant improvement and surpasses the best model that we know by 1 point, showing its outstanding performance. Although our model is not the best on datasets of EN-DE, it is still capable of competing with other outstanding document-level NMT models, like SAN.[21] and QCN [22]. We make the following analysis of the reasons for these listed results:

- Firstly, apposite translation requires more context, while document information is mainly used for semantic disambiguation. Therefore, using the whole document as context, like ACN [9] may perform better. However, after analyzing the translation results, we find that our method which uses word-level automatic routing has more advantages in structured information modeling. Besides, the proposed method is significantly improved on the datasets of TED, which may contains more structured information than the others. See Sect. 5.5 for details.
- Secondly, when we reproduced the experiment of DocT [4], we found that the improvement brought by training strategy is little. Considering the phenomenon above, we do not take the two-step training strategy. But we will keep following it in the future.
- Finally, the context-encoder and the module of *auto-selection* are just updated by the back propagation of the loss between the label and the predicted value. Due to the lack of other supervision, the larger the dataset is, the easier the model overfits the label. Therefore, it can be understood that our method is not significantly improved on the dataset Europarl.

Table 3. Results of ablation study

#	Models	ZH-EN
		TED
1	Transformer [28]	17.11
2	DocT [4]	18.82
3	DocT+AS	19.72
4	Ours (online)	19.50
5	Ours (offline)	20.02

5.4   Ablation Study

We list our results of ablation in Table 3. We mainly produce our ablation study for the following aspects:

Offline vs. Online Document MT. SAN [21] divides the source of context into two cases: *offline* context is both the context of the past and the context of the future; *online* context is only the context of the past. In this part, we compare the result of offline and online document-level MT settings on TED ZH-EN. From the Table 3, we can find that the result of *offline* (row 4) is close to that of *online* (row 5) settings. It is quite self-explanatory that the post sentence as part of context really works in our methods. The proposed method can be extended to the full text as well, but it has achieved impressive performance even if only the previous and the post context sentences are considered, or even only the previous one. Moreover, usually the discourse structure information of local context is usually more meaningful to translation, so we mainly use pre-context and post-context sentences in our experiments.

Universality. According to the experiment of *online* document-level MT, we make an assumption that whether we can apply our methods in other document-level MT methods. To test our intuition, we reproduce the model of DocT [4], whose results are listed in the row 2 and row 3 of Table 3. Compared with the original model, this result achieves an improvements of +0.9 BLEU. The main difference is that our approach of *auto-selection* helps the model to filter some redundant information and focus on the words that are really useful to improve the quality of document-level MT. But we only implement our method on a similar model to ours. We will carry out more experiments in the future to study the universality of our method.

5.5   Analysis

Table 4. Counts of conjunctions

	Ref.	Baseline	DocT	DocT+AS	Ours (online)	Ours (offline)
and	3251	2569	2702	3027	3055	**3210**
but	561	590	587	**606**	594	590
or	233	183	201	**206**	186	197
because	285	295	314	316	307	**337**
so	853	487	497	526	519	**563**
yet	27	9	2	10	**15**	12
then	186	98	84	107	**161**	140

In order to analyze our model's ability of capturing structured information between sentences, we list some common conjunctions that can express the relationship of sentences in Table 4, such as *and, but, because*. According to the statistical results, we find that the document-level MT models tend to generate more conjunctions to capture the structured information. From the comparison of the statistical results of DocT [4] and *online*, we can find that the addition of *auto-selection* allows DocT [4] to add more related words than the original version which is also reflected in the *online* of our model. According to the results of *offline*, we find that with the addition of future context, *offline* tends to add words that express the coordination or causality between sentences.

In order to prove our analysis aforementioned, we list an example in Table 5. The sentences in the source language express both coordination and causality. Among the listed models: baseline, DocT, DocT+AS, online and offline, only the offline model using automatic selection and future context information shows coordination and causality in translation results, which is helpful to prove the effectiveness of our methods.

Table 5. Results of baseline and document-level NMT models

Src: 因为 音乐 可以 帮 他 将 他 的 思维 妄想, 转换 成形 通过 他 的 想象力 和 创造 力 变成 现实 Ref: **because** music allows him to take his thoughts and delusions **and** shape them through his imagination and his creativity , into reality . Baseline: **because** music can help him think of his thinking , transform his imagination through his imagination and creativity .
DocT: **because** music can help him think of his thoughts , change their imagination through his imagination and creativity . DocT+AS: **because** music can help him think of his mind as a delusion , through his imagination and creativity .
Ours(online): **because** music can help him turn his mind into a delusion of his imagination and his creativity . Ours(offline): **because** music can help him think of his delusions , **and** turn it into his imagination and his creativity into reality .

6 Conclusion and Future Work

In this paper, we expand the source of context, and integrate the future context with the sentence to be translated, which is beneficial to the document-level NMT. In order to filter redundant information, we study the routing algorithm in MNMT, and propose a document-level NMT routing algorithm based on this algorithm. With *auto-selection*, the model together with the input is capable of deciding which words to use as context. According to the results of experiments, our *online* model achieves +1.39 BLEU improvement compared with the baseline on TED ZH-EN, which proves the effectiveness of *auto-selection* in document-level NMT; Combined with the future context, our model improves the BLEU by another 0.52 points, which proves that document-level NMT benefits from future contextual information. In addition, we also transplant our method to the previous document-level NMT work, which proves the universality of our method.

We still have a lot of work to do. For example, we do not achieve the expected results on the EN-DE datasets. These problems have already been mentioned above. The lack of other supervision methods and information after the integration of deep coding are the key points that need to be solved in our future work. Besides, we will continue to study the universality of our method in other document-level NMT methods.

Acknowledgments. The authors would like to thank the organizers of CCMT 2021 and the reviewers for their helpful suggestions. This research work is supported by the National Key Research and Development Program of China under Grant No. 2017YFB1002103.

References

1. Cho, K., et al.: Learning phrase representations using RNN encoder-decoder for statistical machine translation. In: Proceedings of the 2014 Conference on Empirical Methods in Natural Language Processing (EMNLP), Doha, Qatar, pp. 1724–1734. Association for Computational Linguistics, October 2014
2. Bahdanau, D., Cho, K., Bengio, Y.: Neural machine translation by jointly learning to align and translate. arXiv preprint arXiv:1409.0473 (2014)
3. Luong, M.-T., Pham, H., Manning, C.D.: Effective approaches to attention-based neural machine translation. In: Proceedings of the 2015 Conference on Empirical Methods in Natural Language Processing, pp. 1412–1421 (2015)
4. Zhang, J., et al.: Improving the transformer translation model with document-level context. In: Proceedings of the 2018 Conference on Empirical Methods in Natural Language Processing, pp. 533–542 (2018)
5. Voita, E., Serdyukov, P., Sennrich, R., Titov, I.: Context-aware neural machine translation learns anaphora resolution. In: Proceedings of the 56th Annual Meeting of the Association for Computational Linguistics (Volume 1: Long Papers), pp. 1264–1274 (2018)
6. Agrawal, R., Turchi, M., Negri, M.: Contextual handling in neural machine translation: look behind, ahead and on both sides. In: 21st Annual Conference of the European Association for Machine Translation, p. 11 (2018)
7. Guillou, L., Hardmeier, C., Lapshinova-Koltunski, E., Loáiciga, S.: A pronoun test suite evaluation of the English-German MT systems at WMT 2018. In: WMT 2018, p. 570 (2018)
8. Voita, E., Sennrich, R., Titov, I.: Context-aware monolingual repair for neural machine translation. In: EMNLP/IJCNLP (1) (2019)
9. Zheng, Z., Yue, X., Huang, S., Chen, J., Birch, A.: Towards making the most of context in neural machine translation. In: IJCAI (2020)
10. Voita, E., Sennrich, R., Titov, I.: When a good translation is wrong in context: Context-aware machine translation improves on deixis, ellipsis, and lexical cohesion. In: Proceedings of the 57th Annual Meeting of the Association for Computational Linguistics, pp. 1198–1212 (2019)
11. Jean, S., Lauly, S., Firat, O., Cho, K.: Does neural machine translation benefit from larger context? arXiv preprint arXiv:1704.05135 (2017)
12. Wang, L., Tu, Z., Way, A., Liu, Q.: Exploiting cross-sentence context for neural machine translation. In: Proceedings of the 2017 Conference on Empirical Methods in Natural Language Processing, pp. 2826–2831 (2017)

13. Miculicich, L., Ram, D., Pappas, N., Henderson, J.: Document-level neural machine translation with hierarchical attention networks. In: Proceedings of the 2018 Conference on Empirical Methods in Natural Language Processing, pp. 2947–2954 (2018)

14. Tu, Z., Liu, Y., Shi, S., Zhang, T.: Learning to remember translation history with a continuous cache. Trans. Assoc. Comput. Linguist. **6**, 407–420 (2018)

15. Kim, Y., Tran, D.T., Ney, H.: When and why is document-level context useful in neural machine translation? In: Proceedings of the Fourth Workshop on Discourse in Machine Translation (DiscoMT 2019), pp. 24–34 (2019)

16. Zhang, B., Bapna, A., Sennrich, R., Firat, O.: Share or not? Learning to schedule language-specific capacity for multilingual translation (2020)

17. Maruf, S., Haffari, G.: Document context neural machine translation with memory networks. In: Proceedings of the 56th Annual Meeting of the Association for Computational Linguistics (Volume 1: Long Papers), pp. 1275–1284 (2018)

18. Tiedemann, J., Scherrer, Y.: Neural machine translation with extended context. In: Proceedings of the Third Workshop on Discourse in Machine Translation, pp. 82–92 (2017)

19. Koehn, P., Knowles, R.: Six challenges for neural machine translation. In: ACL 2017, p. 28 (2017)

20. Sukhbaatar, S., Grave, É., Bojanowski, P., Joulin, A.: Adaptive attention span in transformers. In: Proceedings of the 57th Annual Meeting of the Association for Computational Linguistics, pp. 331–335 (2019)

21. Maruf, S., Martins, A.F., Haffari, G.: Selective attention for context-aware neural machine translation. In: Proceedings of the 2019 Conference of the North American Chapter of the Association for Computational Linguistics: Human Language Technologies (Long and Short Papers), vol. 1, pp. 3092–3102 (2019)

22. Yang, Z., Zhang, J., Meng, F., Gu, S., Feng, Y., Zhou, J.: Enhancing context modeling with a query-guided capsule network for document-level translation. In: Proceedings of the 2019 Conference on Empirical Methods in Natural Language Processing and the 9th International Joint Conference on Natural Language Processing (EMNLP-IJCNLP), pp. 1527–1537 (2019)

23. Li, B., et al.: Does multi-encoder help? a case study on context-aware neural machine translation. In: Proceedings of the 58th Annual Meeting of the Association for Computational Linguistics, pp. 3512–3518 (2020)

24. Cao, Q., Xiong, D.: Encoding gated translation memory into neural machine translation. In: Proceedings of the 2018 Conference on Empirical Methods in Natural Language Processing, pp. 3042–3047 (2018)

25. Kuang, S., Xiong, D.: Fusing recency into neural machine translation with an inter-sentence gate model. In: Proceedings of the 27th International Conference on Computational Linguistics, pp. 607–617 (2018)

26. Jiang, S.: Document-level neural machine translation with inter-sentence attention. arXiv preprint arXiv:1910.14528 (2019)

27. Xiong, H., He, Z., Wu, H., Wang, H.: Modeling coherence for discourse neural machine translation. In: Proceedings of the AAAI Conference on Artificial Intelligence, vol. 33, pp. 7338–7345 (2019)

28. Vaswani, A., et al.: Attention is all you need. In: NIPS (2017)

29. Arivazhagan, N., et al.: Massively multilingual neural machine translation in the wild: findings and challenges. arXiv preprint arXiv:1907.05019 (2019)

30. Cettolo, M., Girardi, C., Federico, M.: WIT3: Web inventory of transcribed and translated talks. In: Proceedings of the 16th Annual Conference of the European Association for Machine Translation, Trento, Italy, 28–30 May 2012, pp. 261–268. European Association for Machine Translation (2012)
31. Koehn, P.: Europarl: a parallel corpus for statistical machine translation. Mt Summit, vol. 5 (2008)
32. Koehn, P., et al.: Moses: open source toolkit for statistical machine translation. In: Proceedings of the 45th Annual Meeting of the Association for Computational Linguistics Companion Volume Proceedings of the Demo and Poster Sessions, Prague, Czech Republic, pp. 177–180. Association for Computational Linguistics, June 2007
33. Sennrich, R., Haddow, B., Birch, A.: Neural machine translation of rare words with subword units. In: Proceedings of the 54th Annual Meeting of the Association for Computational Linguistics (Volume 1: Long Papers), pp. 1715–1725 (2016)
34. Ott, M.: fairseq: a fast, extensible toolkit for sequence modeling. In: Proceedings of the 2019 Conference of the North American Chapter of the Association for Computational Linguistics (Demonstrations), pp. 48–53 (2019)
35. Kingma, D., Ba, J.: Adam: a method for stochastic optimization. Computer Science (2014)
36. Post, M.: A call for clarity in reporting bleu scores. In: Proceedings of the Third Conference on Machine Translation: Research Papers, pp. 186–191 (2018)

Generating Diverse Back-Translations via Constraint Random Decoding

Yiqi Tong[1,2,3], Yidong Chen[1,2(✉)], Guocheng Zhang[1,2], Jiangbin Zheng[1,2,4], Hongkang Zhu[1,2], and Xiaodong Shi[1,2]

[1] Department of Artificial Intelligence, School of Informatics, Xiamen University, Xiamen, China
[2] Key Laboratory of Digital Protection and Intelligent Processing of Intangible Cultural Heritage of Fujian and Taiwan, Ministry of Culture and Tourism, Xiamen, China
[3] Institute of Artificial Intelligence, Beihang University, Beijing, China
[4] AI Lab, School of Engineering, Westlake University, Hangzhou, China
{yqtong,zgccc,hkzhu}@stu.xmu.edu.cn, {ydchen,mandel}@xmu.edu.cn, zhengjiangbin@westlake.edu.cn

Abstract. Back-translation has been proven to be an effective data augmentation method that translates target monolingual data into source-side to improve the performance of Neural Machine Translation (NMT), especially in low-resource scenarios. Previous researches show that diversity of the synthetic source sentences is essential for back-translation. However, the frequently used random methods such as sampling or noised beam search, although can output diverse back-translations, often generate noisy synthetic sentences. To alleviate this problem, we propose a simple but effective constraint random decoding method for back-translation. The proposed method is based on an automatic post-editing (APE) data augment framework, which incorporates fluency boost learning. Moreover, to increase the diversity of synthetic data and ensure quality, we proposed to use an evolution decoding algorithm. Compared with the original back-translation, our method can generate more diverse while less noisy synthetic sentences. The experimental results show that the proposed method can get 0.6 BLEU improvements on the WMT18 EN-DE news dataset and more than 0.4 BLEU improvements on the EN-ZH dataset which is in the medical field, respectively.

Keywords: NMT · Back-translation · Automatic post-editing · Evolution decoding algorithm

1 Introduction

In the past years, attention-based Neural Machine Translation (NMT) has become the mainstream approach because of its significant performance [1,20, 21]. However, to achieve promising performance for a single language pair, millions of parallel sentences are necessary, which are data-hungry in many language pairs. To cope with this issue, researchers investigated using monolingual

© Springer Nature Singapore Pte Ltd. 2021
J. Su and R. Sennrich (Eds.): CCMT 2021, CCIS 1464, pp. 92–104, 2021.
https://doi.org/10.1007/978-981-16-7512-6_8

data for NMT and other natural language processing (NLP) tasks [2]. Specially, [15] proposed back-translation, which makes use of an NMT model with opposite translation direction to translate the target-side monolingual data into the source-side to enrich the parallel training corpus. However, the traditional back-translation still has problems. Current strong NMT model such as Transformer [21] adopts beam search in the decoding stage and generates candidates that only differ with one another by punctuation or minor morphological variations, making the translated sentences lack of diversity [7,12]. On the other hand, the common alternative methods based on random decoding, such as sampling [26], often put too much noise into synthetic sentences, which reduce the data quality.

There are some works attempting to get more diverse and high-quality translation results, e.g. mirror-generative neural machine translation (MGNMT) [29], diverse beam search [22], adding an additional penalization term to expansion the same parent node [12], introducing a discrete latent variables to control generation [7,16], manipulating attention heads [19], etc. While most of these studies have exploited decoding strategy, a few of them have tried developing automatic post-editing (APE) method to efficiently use the monolingual data.

In this work, we proposed a simple but effective constraint random decoding method for back-translation, which follows an APE framework. First, we build fluency boost sentence-pairs by combining the golden source-side sentences and the corresponding pseudo source-side sentences generated by back-translation. Then a sequence-to-sequence APE model was trained to re-generate pseudo source-side sentences, which will be used in the next iterations. Please note that, the above-mentioned process will be iterated several times.

Finally, we build synthetic fluency boost corpus by combining the source-side fluency boost sentences which generated by APE and target-side golden sentences for data augment. During the APE decoding process, a evolution decoding algorithm could be optionally adopted. Our methods can double the training data at maximum and can be applied to any encoder-decoder framework. As far as we know, we are the first to introduce fluency boost learning into the field of back-translation. Experimental results show that the proposed method can get 0.6, 0.4 BLEU improvements over the baseline model on EN-DE, EN-ZH test set.

2   Related Work

The NMT system is known to be extremely data-needed. Previous works proved that the diversity of the training data can provide more discriminative information for the NMT model [5,6]. However, high-quality parallel corpus is limited. To address above problem, [15] proposed back-translation, which utilize abundant amount of mono-lingual data during the model training process. [3,25] broadens the understanding of back-translation and investigated a number of methods like unrestricted sampling, large-scale noised training to generate synthetic source sentences. To explore the actual effects of the back-translation, [14] studied the performance of EN-DE NMT models when incrementally larger amounts of synthetic data are used for training.

Some recent works have looked at the diverse decoding method for NMT. [22] proposed diverse beam search that modifies classical beam search algorithm with a diversity augmented sequence decoding objective and get state-of-the-art results on several language generation tasks at that time. Other than design diversity encouraging decoding algorithm, [7,16] proposed mixture model, which could improve both quality and diversity of the translations by introduced latent variables to control generation. However, this method will increase the difficulty of model training [19]. More recently, to make better use of non-parallel data, [29] proposed a mirror-generative NMT model (MGNMT), which outperforms previous approaches in all investigated scenarios by combing the source-side and target-side monolinguals and corresponding language models organically during the training phase.

Our work was partly followed with [4], which they proposed a fluency boost learning and inference mechanism and get significant improvement over the former Grammar Error Correction (GEC) models. However, they focused on generating more error-corrected data, while we use this strategy to iteratively enhance the predictions of back-translation by rewriting the sentences with our proposed APE model and providing more training signals for NMT model. Moreover, we also incorporated a novel evolution decoding algorithm in the model decoding stage to get more diverse candidates.

3 Proposed Methods

3.1 Fluency Boost Learning

Fluency boost learning (FBL) is an iterative learning strategy, which was first proposed by [4] for solving the GEC problem [9,27]. GEC aims for automatically correcting various types of errors in the given text, while there are mainly Rule-based approaches [17], MT-based approaches [13] and LM-based [18] to solve this problem.

In this paper, we transfer it to the field of back-translation and proposed an Automatic Post-Editing (APE) model which enable to learn how to improve a sentence's diversity and quality without changing its original meaning by FBL. Specifically, the sentences generated by back-translation usually have various errors. Hence we treat it as a MT-based GEC problem, which the source-side is pseudo sentences generated by back-translation and the target-side is golden sentence from parallel corpus. Figure 1 illustrates the training process of our APE model, where PD is parallel dataset, MD is monolingual dataset, NMT and APE are neural machine translation model and automatic post-editing model, respectively. Superscript # denotes the machine translation results, subscript src and trg are source-side sentences and target-side sentences respectively, P stands for it generated by monolingual data. Specifically, We use parallel corpus $PD_{src-trg}$ to training the back-translation model $NMT_{trg-src}$, then we can obtain the fluency boost sentence pairs $PD_{src-src\#}$ by combining MD_{src}^{p} and $MD_{src\#}^{P}$, Where the former is obtained by $PD_{src-trg}$ and the latter is generated from $NMT_{trg-src}$ by decoding MD_{src}^{D}. Finally, we use $PD_{src-src\#}$ to training the NMT-based APE model.

The aim of the NMT model is to maximize the probabilities P of the target languages $\mathbf{Y} = (y_1, ..., y_j)$ given the source language sequences $\mathbf{X} = (x_1, ..., x_i)$, which calculated as follows:

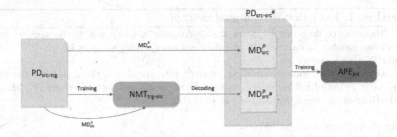

Fig. 1. The training process of our APE model.

$$argmax \frac{1}{N} \sum_{n=1}^{N} log(P_\theta(\mathbf{Y}^n | \mathbf{X}^n)) \tag{1}$$

Where n is the n-th sentence in corpus with a total number of N and θ is model parameters.

In this work, the transformer architecture is used for both NMT and APE models. The difference is, when apply to APE model, the target-side is the sentences which contain various grammatical errors. Our method can be applied to any encoder-decoder framework without any code changes. We expect that other neural sequence-to-sequence based generative model could benefit from our approach, but the choice of the model architecture is not a focus of this paper.

3.2 Evolution Decoding Algorithm

Beam search is a limited-width breadth first search algorithm [11]. For a input sentence x, the generated candidate sequences $\{y_1, ..., y_j\}$ by beam search are highly similar, especially at the beginning part, which is harmful for back-translation or our APE model to generate diversity data. The evolution algorithm [8] is inspired by Darwin's theory, which simulates the natural evolution process of gene sequence and make the next generation of genes stronger through of fittest. When in the decoding stage of back-translation, N-best candidates can be generated like a set of gene sequences. Some words between these candidates are different but have similar semantics, which just provides the basic pre-conditions for our evolution decoding algorithm.

For this part, to further increase the diversity of synthetic data and ensure the quality as much as possible, we proposed an evolution decoding algorithm (EDA), which summarized in Algorithm 1. Formally, we use the offline method[1] to integrate our algorithm into the training process.

As shown in Algorithm 1, EDA selects m-best candidate sequences as the initial population and uses crossover and mutation to modify the original sequence to achieve diversity improvement. For crossover operation, we exchange fragments of two adjacent candidate sequences. If db_i and db_j were chosen, we simply

---

[1] We still use beam search at the model training stage but use EDA at the decoding stage of back-translation or APE.

Algorithm 1. Evolution decoding algorithm

Input: Beam search decoding sequence DB, beam size b, batch size n, the sample size m, maximum number of iterations T and the number of wining samples k

Output: Wining sequence set DB^*

Parameters initial: Population init_pop $= DB^0$, time step $t = 0$, the samples of the wining sequences are initialized as candidate sequences $db_i^* = db_i$ and initial diversity D^0, where D_j^0 is the diversity score of the sample j at time step 0

1: do
2:   for each wining sequence $db_{j_t}^*$ do
3:     $db_{j_{t+1}} \Leftarrow \text{crossover}(db_{j_t}^*)$ # Crossover
4:     $db'_{j_{t+1}} \Leftarrow \text{mutate}(db_{j_{t+1}})$ # Mutation
5:     $D_j^{t+1} \Leftarrow \text{fitness\_function}(db'_{j_{t+1}})$ # Calculate the fitness of sample candidates
6:     $db_{j_{t+1}}^* \Leftarrow \text{select}(D_j^{t+1}, db'_{j_{t+1}})$ # Update winning sequence
7:     $D^{t+1} += \frac{1}{n} D_j^{t+1}$ # Calculate total fitness
8:     $DB^{t+1} \Leftarrow \text{update}(DB^t, db_{j_{t+1}}^*)$ # Update wining sequence set
9:     $t = t + 1$
10: while $D^{t+1} > D^t$ and $t + 1 < T$
11: **return** DB^*

---

split db_i and db_j into s_i^0 and s_i^1, s_j^0 and s_j^1 from the middle position, where superscript 0/1 is the first/second half of the candidate sequence. When using beam search decoding, as we mentioned above, s_i^0 and s_i^1 are usually same, so we set 10% probability to exchange s_i^0 and s_j^0, 90% probability to exchange s_i^1 and s_j^1. Although this may cause the newly generated sequence to be incoherent, previous works [23,24] proved that adding noise to the source-side data can make the model more robust. To avoid getting too much noise, the training data choose 15% of the sequences at random for mutation operation, which follows [2]. If the i-th sequence is chosen, we random replace i-th word with a random word from the vocabulary. We constructed a fitness function to measure sequence diversity, which calculated as follows.

$$d_n = \frac{m \ of \ unique \ n - grams \ in \ k \ translations}{total \ m \ of \ n - grams \ in \ k \ translations} \tag{2}$$

$$u_i = \sum_{i=1}^{k} \frac{unique(s_j) - same(s_j, s_i)}{len(s_j)} \tag{3}$$

$$D_j = \sum_{n=1}^{N} d_j^n + u_j \tag{4}$$

where D_j is our final diversity score for sequence j, which was calculated by d_n and u_i. Specifically, d_n reflects the degree of sub-sequence repeatability for a given sequence, Which higher score means it contains more unique n-gram tuples. We set n to 2 in experiments. However, for a too short sequence, d_n may give an overly high evaluation. To address this shortcoming, we adapt another diversity metric u_j, which measures the difference between j-th sequence and

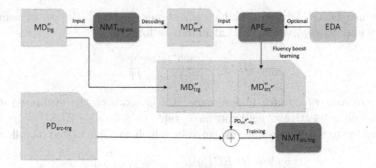

Fig. 2. The whole training process of our NMT model based on FBL and EDA.

others. Where function $unique()$ counts unique words of s_j and $same()$ calculates the identical words between s_j and s_i.

In theory, for a input sequence x, we can get $k-1$ candidates by EDA. After the t-th iteration, $(k-1)2^{(t-1)}+1$ samples will be generated. To reduce the pressure of computation and memory, we only keep the top-k sequences when making the selection operation.

3.3 Joint Training

Figure 2 illustrates the overall architecture of our proposed methods. Where PD is parallel dataset, MD is monolingual dataset, superscript $\#$ and $*$ denotes the synthetic data that generated by NMT and APE respectively, M stands for it comes from bilingual dataset. Our final goal is to get the diversity and high-quality sentence-pairs to improve the performance of NMT model. Therefore, in the first step, $NMT_{trg-src}$ which is trained by golden parallel data $PD_{src-trg}$ to back-translate the target-side monolingual data MD_{trg}^{M} into the source-side pseudo monolingual data $MD_{src\#}^{M}$, then we training the APE_{src} to boost the fluency of $MD_{src\#}^{M}$ and get the fluency-boosted monolingual data $MD_{src\#*}^{M}$. We can carry out multiple rounds of APE to gradually improve the fluency of the corpus. And EDA was applied optional during the APE decoding stage. Finally, we merge $PD_{src-trg}$ and synthetic parallel corpus $PD_{src-trg}^{*}$ to training the $NMT_{src-trg}$.

4 Experimental Setting

4.1 Metrics

To quantitatively assess the quality and diversity of the translation results, we use perplexity to measure the fluency of translation sentences, which a lower perplexity score means the better generalization performance. For evaluate the overall performance of the NMT model, standard BLEU score was calculated.

And we use DEQ (Diversity Enhancement per Quality, [19]) to measure the diversity and quality, which was calculated as follows.

$$DEQ = \frac{(pwb^* - pwb)}{(ref^* - ref)} \tag{5}$$

Where rfb and rfb^* refer to reference BLEU score of the evaluated system and baseline respectively, pwb and pwb^* refer to pair-wise BLEU score of the evaluated system and baseline respectively, which was calculated as follows.

$$pwb = BLEU([y^j], y^k)_{j \in [k], k \in [k], j \neq k} \tag{6}$$

Where $\{y^1, y^2, ..., y^k\}$ are k translation hypotheses of a source sentence x. Lower pwb and higher rfb means better results.

4.2   Dataset

We evaluate NMT training on parallel corpus and with additional monolingual data, which consist of the following five parts.

2M EN-DE. We randomly select 2M sentence-pairs in the news filed from WMT18 for English-German translation task.

80K EN-DE. To simulate low-resource scenarios, we randomly select 80K sentence-pairs from 2M EN-DE.

2M EN-ZH. We randomly select 2M sentence-pairs in the medical field from 10M English-Chinese which collected by our own.

2M DE. Contain 2M German monolingual sentences from News Crawl.

2M ZH. Contain 2M Chinese monolingual sentences from 10M EN-ZH, which the 2M ZH training data has been excluded.

Finally, we choose newstest2013-2018 and randomly select 3K from 10M EN-ZH as our test set for EN-DE task and EN-ZH task respectively, which 2M EN-ZH training data has already been excluded.

4.3   Experiment Settings

We use the Moses tokenizer [10] and learn a joint source and target Byte-Pair-Encoding [15] by fastBPE[1] with 35K types. Before we conduct the random selection, all sentences were lowercased, and which length longer than 150 sub-words were removed. We also remove the sentence pairs whose length ration exceed 1.5 between the source-side and the target-side. The hyper-parameters for our neural NMT and APE model are adopt from [28]. All models are trained on NVIDIA GeForce RTX 2080Ti GPUs and use label smoothing with a uniform prior distribution over the vocabulary $\epsilon = 0.1$. We use same hyper-parameters for all experiments.

---

[1] https://github.com/glample/fastBPE.

Table 1. Under different data scenarios, model performance comparison.

80k bilingual training data and with 2M monolingual data

Models	NST13	NST14	NST15	NST16	NST17	NST18	AVG
BITEXT	15.25	14.32	16.51	19.09	15.99	21.05	17.03
BITEXT+BEAM	21.47	22.26	23.85	28.41	23.52	32.19	25.28
BITEXT+SAMPLING	21.38	21.85	23.86	28.81	23.21	32.16	25.21
BITEXT+APE1	21.44	22.48	**24.56**	**29.43**	23.73	32.89	25.76
BITEXT+APE2	21.51	22.71	24.27	29.06	23.73	**33.57**	25.81
BITEXT+APE3	**21.66**	**22.86**	24.55	29.27	**23.78**	33.31	**25.90**

2M bilingual training data and with 2M monolingual data

Models	NST13	NST14	NST15	NST16	NST17	NST18	AVG
BITEXT	24.64	25.83	27.93	33.82	27.49	39.76	29.91
BITEXT+BEAM	25.94	27.76	29.6	35.77	28.76	42.12	31.66
BITEXT+SAMPLING	25.58	27.59	29.86	35.85	28.91	42.35	31.69
BITEXT+APE1	**26.15**	**28.41**	**30.22**	36.17	29.11	42.61	32.11
BITEXT+APE2	26.13	27.98	30.05	**36.31**	**29.16**	43.2	**32.14**
BITEXT+APE3	25.95	28.15	30.06	36.12	28.87	**42.75**	31.98

Table 2. The comparison of different data augment methods.

Methods	perplexity	pwb	|DEQ|
Baseline	478.00	75.68	0
Back-translation	392.11	80.01	1.89
APE	**324.35**	**73.02**	**3.11**

5   Results and Analysis

5.1   Main Results

As shown in Table 1, We conducted experiments on two different data scale. Where BITEXT is baseline NMT system without adopting any data augment methods. +BEAM and +SAMPLING are NMT systems with standard back-translation, which adopts different decoding strategies. +APE are our proposed APE model with different iterations. As the APE rounds increase, the BLEU score shows an overall upward trend, which is consistent with our assumption, with the iteration of APE, there will be more higher-quality candidate translations to choose from. Compared with +BEAM, our best model can achieve average 0.60 and 0.48 BLEU score improvement respectively through APE. These results suggest that our method can improve the quality of back-translation. Moreover, insufficient NMT model training leads to the poor-quality of back-translation. So our method is more effective in low-resource scenarios, which the improvement of 80K BITEXT+APE is larger than 2M BITEXT+APE.

Table 3. Model performance comparison between different decoding strategies.

Decoding Strategy	NST13	NST14	NST15	NST16	NST17	NST18	AVG
BEAM	25.94	27.76	29.6	35.77	28.76	42.12	31.66
SAMPLING	25.58	27.59	29.86	35.85	28.91	42.35	31.69
EDA_N-GRAM	25.86	**28.14**	29.96	36.05	**29.14**	42.54	31.95
EDA_DIFF	25.54	27.3	29.98	35.74	28.88	42.31	31.63
EDA	25.72	27.51	30.1	35.94	29.01	42.6	31.81
APE+EDA	**26.02**	28.01	**30.15**	**36.14**	28.83	**43.02**	**32.03**

Table 4. Translation diversity and quality comparison.

Decoding strategy	perplexity	pwb
Beam search	364.06	80.01
SAMPLING	1138.28	12.93
EDA	**418.78**	**74.79**

5.2 Quantitative Analysis

Furthermore, we want to prove that our proposed method can both improve diversity and quality of the synthetic data. For this purpose, we conducted analysis experiments on the translations and use perplexity, Pair-wise BLEU (pwb) and DEQ as evaluation metrics.

Specifically, to evaluate the APE model at corpus level, we randomly select 7M monolingual data from News Crawl to training a 5-gram language model, then use it to calculate the average perplexity. And we select 5-best candidates for each source sentences in newstest2013-2018 to calculate pwb and DEQ score. As shown in Table 2, through our proposed fluency boost learning method, the quality and diversity of synthetic data are significantly improved. Compared with the baseline system BITEXT, the perplexity and pwb dropped by 153.65 and 2.66 through APE, which indicating that both the diversity and quality of the translations have been improved. On the contrary, back-translation will both reduce the diversity and quality of the translations, which perplexity and pwb increased 67.76 and 1.22 respectively in our experiment. Finally, compared with back-translation, DEQ increased by 1.22, which also proves the diversity and quality are both improved by APE.

To further boost synthetic data diversity and explore the effectiveness of EDA, we conduct experiments to compare the performance of the NMT model with different decoding strategies. As shown in Table 3, all models are trained with 2M EN-DE and adopt 2M DE for data augment. Where EDA_DIFF adopts formula (2) as fitness function, EDA_GRAM and EDA adopts both formula (3) and (4) as fitness function, respectively. The average overall score of EDA is slightly higher than BEAM and SAMPLING system. With one round of fluency boosting, APE+EDA model achieved best performance. As mentioned in

Table 5. Model performance on EN-ZH test set.

Model	Test set
BITEXT	37.23
BITEXT+BEAM	38.51
BITEXT+SAMPLING	37.98
BITEXT+APE1	**38.93**

Table 6. Case study.

Example 1	
Src	doch in amerika wird mehr so viel getanktwie früher
Ref	americans don't buy as much as gas as they used to
Baseline	but in America *is* not a lot more than is before
Ours	but in America, ***however***, is not much more the before
Example 2	
Src	30 vorschläge standen zur auswah, fünf sind noch im rennen
Ref	there were 30 proposals to choose from, five of which are still in the running
Baseline	30 proposals were made ***to selection***, five are still in the race
Ours	***thirty*** proposals were made ***to select***, five are still in the race

formula (2) and (3), we have defined two indicators to measure diversity for candidates re-ranking, so we perform ablation experiments to test the effects of the two indicators separately. The experimental result in Table 3 shows that the BLEU of EA_N-GRAM system is 0.29 higher than BEAM, indicating that it is feasible to use the diversity of sub-sequences to measure the diversity of whole sequence. On the other hand, EA_DIFF can also produce equivalent results to the standard beam search, but the effect is not as good as the EDA or EA_N-GRAM.

We also did a quantitative analysis of EDA, as shown in Table 4, our evolutionary decoding algorithm can achieve a compromise between beam search and sampling. Compared with beam search, EDA improves diversity of generated data. And compared with the sampling, EDA introduces less noise.

5.3   Qualitative Analysis

For testing the applicability of our proposed model in other domains, we conducted experiments on 2M EN-ZH medical data and with 2M ZH monolingual data. As shown in Table 5, we can get conclusions similar to 2M EN-DE experiments, BITEXT+APE1 still get best performance. But the improvement brought by data augmentation is limited. We believe that the quality of 2M EN-

ZH is higher than 2M EN-DE. So improvement brought by data augmentation is limited.

To observe the effect more intuitively, we give two examples to illustrate the improvement brought by our model. As shown in Table 6, all models are trained with 80K EN-DE and use 2M DE for data augment. Compared with baseline model, our model could not only correct word errors and grammatical errors like "is" in Example 1 but also improved sentences diversity like "however" in Example 1 and "thirty" in Example 2, which proved that our method can both improve the diversity and quality of the back-translation sentence by introducing fluency boost learning. However, due to the NMT model did not correctly translate "getankt" into "gas", our model was not corrected it either.

6   Conclusion

To promote the diversity and quality of synthetic data generated by back-translation, in this paper, we proposed a fluency boost learning based data augment framework, which could extend the origin corpus and applied to any sequence to sequence machine translation model. Furthermore, we performed experiments on different language pairs and resource scenarios to prove our methods could boost both the quality and diversity of the synthetic corpus generated by back-translation. Finally, the experiment results on EN-DE and EN-ZH showed that our proposed methods were effective. In future work, we will explore the influence of noise bring by back-translation under different data scales and further improve our evolution decoding algorithm.

Acknowledgements. The authors would like to thank the three anonymous reviewers for their comments on this paper. This research was supported in part by the National Natural Science Foundation of China under Grant Nos. 62076211, U1908216 and 61573294, and the Outstanding Achievement Late Fund of the State Language Commission of China under Grant WT135-38.

References

1. Bahdanau, D., Cho, K., Bengio, Y.: Neural machine translation by jointly learning to align and translate. arXiv preprint arXiv:1409.0473 (2014)
2. Devlin, J., Chang, M.W., Lee, K., Toutanova, K.: Bert: pre-training of deep bidirectional transformers for language understanding. arXiv preprint arXiv:1810.04805 (2018)
3. Edunov, S., Ott, M., Auli, M., Grangier, D.: Understanding back-translation at scale. arXiv preprint arXiv:1808.09381 (2018)
4. Ge, T., Wei, F., Zhou, M.: Fluency boost learning and inference for neural grammatical error correction. In: Proceedings of the 56th Annual Meeting of the Association for Computational Linguistics (Volume 1: Long Papers), pp. 1055–1065 (2018)
5. Gimpel, K., Batra, D., Dyer, C., Shakhnarovich, G.: A systematic exploration of diversity in machine translation. In: Proceedings of the 2013 Conference on Empirical Methods in Natural Language Processing, pp. 1100–1111 (2013)

6. Gong, Z., Zhong, P., Hu, W.: Diversity in machine learning. IEEE Access **7**, 64323–64350 (2019)
7. He, X., Haffari, G., Norouzi, M.: Sequence to sequence mixture model for diverse machine translation. arXiv preprint arXiv:1810.07391 (2018)
8. Hinterding, R., Michalewicz, Z., Eiben, A.E.: Adaptation in evolutionary computation: a survey. In: Proceedings of 1997 IEEE International Conference on Evolutionary Computation (ICEC 1997), pp. 65–69. IEEE (1997)
9. Junczys-Dowmunt, M., Grundkiewicz, R., Guha, S., Heafield, K.: Approaching neural grammatical error correction as a low-resource machine translation task. arXiv preprint arXiv:1804.05940 (2018)
10. Koehn, P., et al.: Moses: open source toolkit for statistical machine translation. In: Proceedings of the 45th Annual Meeting of the ACL on Interactive Poster and Demonstration Sessions, pp. 177–180. Association for Computational Linguistics (2007)
11. Kool, W., Van Hoof, H., Welling, M.: Stochastic beams and where to find them: the gumbel-top-k trick for sampling sequences without replacement. arXiv preprint arXiv:1903.06059 (2019)
12. Li, J., Monroe, W., Jurafsky, D.: A simple, fast diverse decoding algorithm for neural generation. arXiv preprint arXiv:1611.08562 (2016)
13. Malmi, E., Krause, S., Rothe, S., Mirylenka, D., Severyn, A.: Encode, tag, realize: high-precision text editing. arXiv preprint arXiv:1909.01187 (2019)
14. Poncelas, A., Shterionov, D., Way, A., de Buy Wenniger, G.M., Passban, P.: Investigating backtranslation in neural machine translation (2018)
15. Sennrich, R., Haddow, B., Birch, A.: Improving neural machine translation models with monolingual data. arXiv preprint arXiv:1511.06709 (2015)
16. Shen, T., Ott, M., Auli, M., Ranzato, M.: Mixture models for diverse machine translation: tricks of the trade. arXiv preprint arXiv:1902.07816 (2019)
17. Sidorov, G., Gupta, A., Tozer, M., Catala, D., Catena, A., Fuentes, S.: Rule-based system for automatic grammar correction using syntactic n-grams for english language learning (l2). In: Proceedings of the Seventeenth Conference on Computational Natural Language Learning: Shared Task, pp. 96–101 (2013)
18. Stahlberg, F., Bryant, C., Byrne, B.: Neural grammatical error correction with finite state transducers. arXiv preprint arXiv:1903.10625 (2019)
19. Sun, Z., Huang, S., Wei, H.R., Dai, X., Chen, J.: Generating diverse translation by manipulating multi-head attention. In: AAAI, pp. 8976–8983 (2020)
20. Tong, Y., Zheng, J., Zhu, H., Chen, Y., Shi, X.: A document-level neural machine translation model with dynamic caching guided by theme-rheme information. In: Proceedings of the 28th International Conference on Computational Linguistics, pp. 4385–4395 (2020)
21. Vaswani, A., et al.: Attention is all you need. In: Advances in Neural Information Processing Systems, pp. 5998–6008 (2017)
22. Vijayakumar, A.K., et al.: Diverse beam search for improved description of complex scenes. In: Thirty-Second AAAI Conference on Artificial Intelligence (2018)
23. Vincent, P., Larochelle, H., Bengio, Y., Manzagol, P.A.: Extracting and composing robust features with denoising autoencoders. In: Proceedings of the 25th International Conference on Machine Learning, pp. 1096–1103 (2008)
24. Vincent, P., Larochelle, H., Lajoie, I., Bengio, Y., Manzagol, P.A., Bottou, L.: Stacked denoising autoencoders: Learning useful representations in a deep network with a local denoising criterion. J. Mach. Learn. Res. **11**(12), 3371–3408 (2010)

25. Wu, L., Wang, Y., Xia, Y., Tao, Q., Lai, J., Liu, T.Y.: Exploiting monolingual data at scale for neural machine translation. In: Proceedings of the 2019 Conference on Empirical Methods in Natural Language Processing and the 9th International Joint Conference on Natural Language Processing (EMNLP-IJCNLP), pp. 4198–4207 (2019)
26. Xu, W., Niu, X., Carpuat, M.: Differentiable sampling with flexible reference word order for neural machine translation. arXiv preprint arXiv:1904.04079 (2019)
27. Yannakoudakis, H., Rei, M., Andersen, Ø.E., Yuan, Z.: Neural sequence-labelling models for grammatical error correction. In: Proceedings of the 2017 Conference on Empirical Methods in Natural Language Processing, pp. 2795–2806 (2017)
28. Zhang, J., et al.: THUMT: an open source toolkit for neural machine translation. arXiv preprint arXiv:1706.06415 (2017)
29. Zheng, Z., Zhou, H., Huang, S., Li, L., Dai, X.Y., Chen, J.: Mirror-generative neural machine translation. In: International Conference on Learning Representations (2019)

ISTIC's Neural Machine Translation System for CCMT' 2021

Hangcheng Guo, Wenbin Liu, Yanqing He[⊠], Zhenfeng Wu, You Pan,
Tian Lan, and Hongjiao Xu

Research Center of Information Theory and Methodology,
Institute of Scientific and Technical Information of China, Beijing 100038, China
{guohc2020,liuwb2019,heyq,wuzf,pany,lantian,xuhj}@istic.ac.cn

Abstract. This paper demonstrates an overview and the technical details of the neural machine translation system developed by the Institute of Scientific and Technical Information of China (ISTIC) for the 17th China Conference on Machine Translation (CCMT' 2021). ISTIC participated in the following four machine translation (MT) evaluation tasks: MT task of Mongolian-to-Chinese daily expressions, MT task of Tibetan-to-Chinese government documents, MT task of Uyghur-to-Chinese news, and MT task of Russian-to-Chinese in low resource languages. Our system is based on Transformer architecture and several effective strategies are adopted to improve the quality of translation, such as corpus filtering, back translation, data augmentation, context-based system combination, model averaging, model ensemble, and reranking. The paper presents the system performance under different parameter settings.

Keywords: Neural machine translation · Self-attention mechanism · Context-based system combination

1 Introduction

The machine translation team of the Institute of Scientific and Technical Information of China (ISTIC) participated in four machine translation evaluation tasks in the 17th China Conference on Machine Translation (CCMT'2021), including three bilingual evaluation tasks (Mongolian-to-Chinese daily expressions track, namely M2C; Tibetan-to-Chinese government documents track, namely T2C; Uyghur-to-Chinese news track, namely U2C) and one low resource evaluation task (Russian-to-Chinese tourism oral track, namely R2C). This paper describes the general overview and technical details of ISTIC's neural machine translation system for CCMT' 2021.

In this evaluation, we adopted the neural machine translation architecture of Google Transformer [1] as the basis of our system. As regards data source, the monolingual data released by the evaluation organizer is filtered to construct

This work was supported by ISTIC Fund (ZD2021-17).

J. Su and R. Sennrich (Eds.): CCMT 2021, CCIS 1464, pp. 105–116, 2021.
https://doi.org/10.1007/978-981-16-7512-6_9

pseudo parallel corpus through the back-translation method in M2C, T2C, and U2C evaluation tasks; the pseudo parallel corpus and the original given bilingual parallel corpus are used together as the training set of our neural machine translation system. External data of self-built Russian-Chinese dictionary and bilingual parallel corpus are introduced in the R2C evaluation task since the scale of given data is too small. In terms of data pre-processing, we proposed a general pre-processing method and a specific pre-processing method for the given data. Several filtering methods of the corpus are explored to reduce the data noise and improve the data quality. As for model construction, the context-based system combination method inputs the source sentence and its translation results from multiple machine translation systems as additional signals into multi-encoders respectively, which are weighted by attention mechanism to get combination result by encoder combination and decoder combination through gate mechanism. Model averaging and model ensemble strategies are adopted to generate the final output translation. We removed spaces between words and restored the target language translation results to the prescribed XML format in data post-processing. For each task, we compared the system performance under different parameter settings and further analyzed the experimental results.

The structure of this paper is as follows: the second part introduces the technical architecture of ISTIC's neural machine translation system; the third part introduces methods used in different tasks; the fourth part introduces the parameter settings, data pre-processing, experimental results, and related analysis; the fifth part gives the conclusion and future work.

2   System Architecture

Figure 1 shows the overall flow chart of our neural machine translation system in this evaluation which includes data pre-processing, model training, model decoding, and data post-processing (see Fig. 1).

2.1   Baseline System

Our baseline system used in participated evaluation tasks is Transformer, which includes an encoder and a decoder (see Fig. 2). The transformer is completely based on an attention mechanism. It can achieve algorithm parallelism, speed up model training, further alleviate long-distance dependence and improve translation quality [2].

The encoder and decoder are formed by stacking n identical layer blocks, where n is set to 6. Each layer of encoder contains two sub-modules, namely a multi-head self-attention module and a feed-forward neural network module. The multi-head self-attention module divides the dimension of the hidden state into multiple parts, and each part is separately calculated by using the self-attention function. Furthermore, these output vectors are concatenated together. The multi-head mechanism enables the model to pay more attention to the feature information of different positions and different sub-spaces. The multi-head attention method includes two steps: 1) dot product attention calculation; 2) multi-head attention calculation. The calculation method of dot product attention can be expressed as:

$$Attention(Q, K, V) = softmax(\frac{QK^T}{\sqrt{d_k}}) \cdot V \tag{1}$$

where Q is the query vector, K is the key vector, V is the value vector, and d_k is the dimension of the hidden layer state. Based on dot product attention, the calculation method of the multi-head attention mechanism can be expressed as:

$$MutiHead(Q, K, V) = Concat(head_i, ..., head_n) \cdot W^O \tag{2}$$

where W^O is the matrix parameter. The attention value of each head is:

$$head_i = Attention(Q \cdot W_i^Q, K \cdot W_i^K, V \cdot W_i^V) \tag{3}$$

Each layer of the decoder is composed of three sub-modules. In addition to the two modules similar to the encoder, a decoder-encoder attention module is

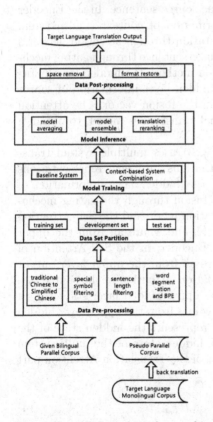

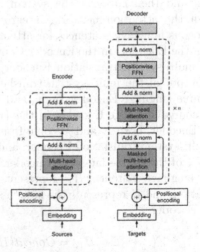

Fig. 1. Overall flow chart for machine translation tasks.

Fig. 2. Transformer model structure.

added between them and can focus attention on source language information in the decoding process. To avoid the problem that too many layers cause the model to be difficult to converge, both the encoder and the decoder use residual connection and hierarchical regularization techniques. To make the model better obtain the position information of the input sentence, additional position encoding vectors are added to the input layer of the encoder and decoder. After the encoder obtains a hidden state, the Transformer model inputs the hidden state into the softmax layer and scores with candidate vocabulary to obtain the final translation result.

2.2 Our System

Based on the transformer model, we propose a context-based [3] system combination method, which also adopts an encoder-decoder structure composed of n identical network layers, where n is set to 6. Two different methods of system combination are designed according to the fusion in different positions, which are Encoder Combination method and Decoder Combination method. Both of them adopt multi-encoder [4] to encode the source sentences and the context information from machine translation results of the source sentence. In the Encoder Combination method, the hidden layer information of context (multi-system translation) is transformed into new representation through attention network, and merges the hidden layer information of source sentence through gating mechanism at encoder end; In Decoder Combination method, the hidden layer information of multi-system translation and the hidden layer information of source sentence is calculated at the decoder to obtain the fusion vector. The attention calculation method is the same as the original transformer model, to obtain a higher quality fusion translation.

The Encoder Combination model (see Fig. 3) uses multiple system translations, and then converts the system translations into new representations through the attention network, integrating the hidden layer information of homologous language sentences for attention fusion through the gating mechanism in the Encoder. In the Encoder Combination mode and the Self-Attention of the multi-system translation Encoder, Q, K, and V are all from the upper layer output of the multi-system translation Encoder; in the Self-Attention of the source language Encoder, Q, K, and V are all from the upper layer output of the source language Encoder; in the Translation Attention of the source language Encoder, both K and V come from the upper hidden layer state H_{T_r} of the multi-system translation Encoder, and Q comes from the upper layer hidden state H_s of the source language Encoder. H_s represents the hidden state of the source language sentence, H_{T_r} represents the hidden state of the multi-system translation, and H represents the hidden state of the Translation Attention part of the Encoder.

$$H_{T_r} = Concat(H_{T_r1}, ..., H_{T_rn}) \qquad (4)$$

$$H = MutiHead(H_{T_r}, H_s) \qquad (5)$$

The Decoder Combination model (see Fig. 4) combines the hidden layer information of multiple encoders with attention in the decoder. The Decoder can process multiple encoders separately, and then fuse them using the gating mechanism inside the Decoder to obtain the combined vector. In the Decoder Combination mode and the Self-Attention of the target language Decoder, Q, K, and V are all from the output of the previous layer of the target language Decoder; in the Translation Attention of the target language Decoder, Q comes from the output of the upper layer of the target language Decoder, K comes from the upper hidden layer state H_s of the source language Encoder, and V comes from the upper hidden layer state H_{T_r} of the multi-system translation Encoder; in the Encoder-Decoder Attention of the target language Decoder, Q comes from the upper layer output of the target language Decoder, K, V come from the previous output of the source language Encoder. H_s represents the hidden layer state of the source language sentence, H_{T_r} represents the hidden layer state of the multi-system translation, $H_{Decoder}$ represents the hidden layer state of the upper layer output of the Decoder, and H represents the hidden state of the Translation Attention part of the Decoder.

$$H = MutiHead(H_{T_r}, H_s, H_{Decoder}) \tag{6}$$

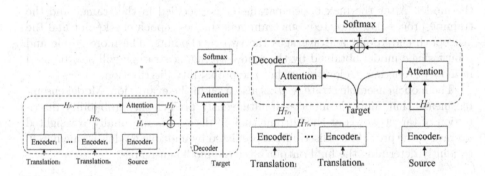

Fig. 3. Encoder combination model.   **Fig. 4.** Decoder combination model.

3   Methods in Different Tasks

In this evaluation, ISTIC participated in the four tasks of the Mongolia-Chinese, Tibetan-Chinese, Uyghur-Chinese, and Russian-Chinese. The methods used in each task are introduced below. For convenience, M2C represents Mongolian-to-Chinese daily expressions MT task, U2C represents Uyghur-to-Chinese news field MT task, T2C represents Tibetan-to-Chinese government documents MT task, and R2C represents Russian-to-Chinese low resource languages MT task.

3.1 M2C Task, U2C Task, and T2C Task

In the M2C task, U2C task, and T2C task, the parallel corpus is small in scale, so the back translation method is used to construct a pseudo parallel corpus. We used the given parallel corpus data of Mongolian-Chinese, Uyghur-Chinese, and Tibetan-Chinese released by the evaluation organizer to train the Chinese-to-Mongolian, Chinese-to-Uyghur, and Chinese-to-Tibetan translation models. The Chinese monolingual data filtered by Elasticsearch [5] is translated into the pseudo parallel corpus of Mongolian-Chinese, Uyghur-Chinese, and Tibetan-Chinese as a supplement to the training set of neural machine translation model.

In model training, the Transformer model based on the self-attention mechanism is adopted as the baseline model, and the Encoder Combination and Decoder Combination are introduced respectively. The source sentence and context information are integrated into the combination system. Here, the source language sentence, the target language sentence, and the machine translation of the source language sentence through the baseline model are used as the context information respectively.

The model averaging strategy [6] is used in the evaluation. Model averaging means that the parameters of the same model at different training moments are averaged and the more robust model parameters are obtained, which is helpful to reduce the instability of model parameters and enhance the robustness of the model. After the max epoch parameter is specified in the trainer and the training process is completed, our team gets the best epoch checkpoint and the last epoch checkpoint and averages the two checkpoints. The more stable and robust single model obtained by the model averaging strategy will also be used in model averaging, to jointly predict the probability distribution.

The model ensemble strategy [7] is also used in the evaluation. Model ensemble means that multiple models simultaneously predict the probability distribution of target words at the current time in decoding, and finally, a weighted average of the probability distribution predicted by multiple models is calculated, to jointly determine the final output after the model ensemble.

3.2 R2C Low Resource Task

The success of neural machine translation is closely related to computing resources, algorithm models, and data resources, especially the scale of bilingual training data. In the R2C task, the number of sentence pairs of parallel corpus available for training is as low as 50,000. Therefore, the introduction of external resources can effectively improve the performance of the machine translation system. Here, 123,605 phrase pairs and 55,504 sentence pairs from a self-built Russian-Chinese dictionary are used.

In our constrained system, the Encoder Combination and Decoder Combination are also adopted based on the Transformer model. Here, the target sentences in the training corpus are directly used as the context of the source sentences for system combination training. The strategy of model ensemble is also used. In our unconstrained system, 123,605 phrase pairs and 55,504 sentence pairs from a self-built Russian-Chinese dictionary are used as the training set together with the training corpus released by the evaluation organizer.

4   Experiments

4.1   System Settings

The open-source project fairseq [8,9] is chosen for this evaluation system. The main parameters are set as follows. Each model uses 1–3 GPUs for training, and the batch size is 2048. The embedding size and hidden size are set to 1024, the dimension of the feed-forward layer is 4096. We use six self-attention layers for both encoder and decoder, and the multi-head self-attention mechanism has 16 heads. The dropout mechanism [10] was adopted, and dropout probabilities are set to 0.3. BPE [11] is used in all experiments, where the merge operations is set to 32000. The maximum number of tokens is set to 4096. The loss function is set to *"label_smoothed_cross_entropy"*. The parameter *adam_betas* is set to (0.9, 0.997). For the baseline system, the initial learning rate is 0.0007, the warm-up steps are set to 4000, and the maximum epoch number is set to 30. For the Encoder Combination system and Decoder Combination system, the initial learning rate is 0.0001, the warm-up steps are set to 4000, and the maximum epoch number is set to 10.

4.2   Data Pre-processing

In the M2C task, U2C task, and T2C task, the bilingual parallel corpus, and monolingual corpus are released by the evaluation organizer. In the R2C low resource task, the only bilingual parallel corpus is released.

1. Bilingual Parallel Corpus Pre-processing.
The Mongolian-Chinese parallel corpus is in the field of daily expressions, the Tibetan-Chinese parallel corpus is in the field of government literature, the Uygur-Chinese parallel corpus is in the field of news, and the Russian-Chinese parallel corpus belongs to low resource languages. The characteristics of language pairs of the evaluation tasks are similar as well as different. Therefore, a two-stage pre-processing method is designed as a general pre-processing stage and a specific pre-processing stage [12].

The the general pre-processing stage includes conversion from traditional Chinese to simplified Chinese, conversion between full angle and half-angle, special character filtering, same content filtering, sentence length filtering, and sentence length ratio filtering. Among them, sentence length of the Chinese language is calculated in the unit of "character" and sentence length of non-Chinese language is calculated in the unit of "token". Sentence length filtering removes sentence pairs which source sentence length or target sentence length exceeds the range of [1, 200]. Sentence length ratio filtering excludes the sentence pairs whose ratio of source sentence length and target sentence length exceeds the range of [0.1, 10]. In the specific pre-processing stage, Chinese word segmentation is implemented using the lexical tool Urheen [13] and Chinese word segmentation is implemented using the lexical tool Polyglot [14].

In the M2C task, U2C task, and T2C task, parallel corpus data other than "imu-test-mnzh-cwmt2018", "imu-test-uyzh-cwmt2018" and "imu-test-tizh-cwmt2018" are taken as training sets. All the training set is processed in two stages. In the pre-processing of the development set and test set, the same content filtering, sentence length filtering, and sentence length ratio filtering are excluded. In the R2C task, all parallel corpus data is taken as the training set. Same content filtering, sentence length filtering, and sentence length ratio filtering are excluded in training data, development set, and test set. The number of sentence pairs of training set before and after data pre-processing is shown in Table 1.

Table 1. Training set data reprocessing results.

Direction	Before Pre-processing	After Pre-processing
mn-ch	269462	249069
uy-ch	170061	165143
ti-ch	162096	153324
ru-ch	50000	50000

2. Monolingual Corpus Pre-processing.

In the M2C task, U2C task, and T2C task, the scale of the Chinese monolingual corpus released by the CCMT'2021 evaluation organizer is 662904 news articles, about 11 million words. The monolingual data is filtered and screened by the Elasticsearch retrieval tool [12], and then both general pre-processing and specific pre-processing are used to obtain the final monolingual data for back-translation. The quantity is shown in Table 2.

For the selected monolingual data, this evaluation adopts a back-translation strategy to construct pseudo parallel corpus to enhance the machine translation results. According to the parallel corpus of Mongolian-Chinese, Uyghur-Chinese, and Tibetan-Chinese provided by CCMT' 2021, the neural machine translation model of Chinese-to-Mongolian, Chinese-to-Uyghur and Chinese-to-Tibetan are constructed, and then the selected Chinese monolingual corpus is translated into

Table 2. The scale of pseudo parallel corpus.

Direction	Sentence scale
mn-ch	240000
uy-ch	159894
ti-ch	140000

the corresponding minority languages through these models. Finally, the pseudo parallel corpus obtained from back translation and the preprocessed high-quality bilingual parallel corpus provided by CCMT' 2021 are mixed for training, to improve the machine translation quality of Mongolian Chinese, Tibetan Chinese, and Uighur Chinese.

4.3 Experimental Results

In the Mongolian-to-Chinese translation evaluation task, the primary system (mc-2021-istic-primary-a) trains 10 epochs with the Encoder Combination model system and uses the last epoch checkpoint to decode. The contrast system 1 (mc-2021-istic-contrast-b) trains 10 epochs with the Encoder Combination model system and uses the model ensembling strategy to decode. The contrast system 2 (mc-2021-istic-contrast-c) trains 10 epochs with a Decoder Combination model system and uses the last epoch checkpoint to decode. The above three systems take the translated sentences of the source language sentences decoded by the intermediate translation model as the context and use the pseudo-parallel corpus constructed from monolingual data as the supplement of the training set. The BLEU5-SBP [15] scoring results on the released test set are shown in Table 3. Among all constrained systems, the primary system (mc-2021-istic-primary-a) ranked third. Among all of the participated systems, mc-2021-istic-contrast-b ranked fifth, mc-2021-istic-primary-a ranked sixth, and mc-2021-istic-contrast-c ranked seventh.

Table 3. BLEU5-SBP scoring of Mongolian-to-Chinese track on released test set.

System	BLEU5-SBP
mc-2021-istic-primary-a (encoder combination model)	0.3566
mc-2021-istic-contrast-b (encoder combination + model ensembling)	0.3607
mc-2021-istic-contrast-c (decoder combination model)	0.354

In the Tibetan-to-Chinese translation evaluation task, the primary system (tc-2021-istic-primary-a) takes the source language sentences as the context, trains 30 epochs with the Transformer baseline system, and uses the last epoch checkpoint to decode. The contact system 1 (tc-2021-istic-contact-b) uses the source language sentences as context, trains 30 epochs with the Transformer baseline system, and uses the model ensembling strategy to decode. The contrast system 2 (tc-2021-istic-contact-c) takes the target language sentence as the context, trains 10 epochs with the Encoder Combination model system and uses the last epoch checkpoint to decode. The above three systems do not use monolingual data, and the BLEU5-SBP scoring results on the released test set are shown in Table 4. Among all constrained systems, the primary system (tc-2021-istic-primary-a) ranked third. Among all of the participated systems, tc-2021-istic-contrast-c ranked fourth, tc-2021-istic-contrast-b ranked fifth, and tc-2021-istic-primary-a ranked the sixth.

Table 4. BLEU5-SBP scoring of Tibetan-to-Chinese track on released test set.

System	BLEU5-SBP
tc-2021-istic-primary-a (baseline)	0.1567
tc-2021-istic-contrast-b (baseline + model ensembling)	0.1678
tc-2021-istic-contrast-c (encoder combination model)	0.1737

In the Uyghur-to-Chinese translation evaluation task, the primary system (uc-2021-istic-primary-a) trains 10 epochs with a Decoder Combination model system and uses the best epoch checkpoint to decode. The contrast system 1 (uc-2021-istic-contrast-b) trains 30 epochs with a Transformer baseline system and uses the model ensembling strategy to decode. The contrast system 2 (uc-2021-istic-contrast-c) trains 10 epochs with the Encoder Combination model system and uses the model ensembling strategy to decode. The above three systems take the translated sentences of the source language sentences decoded by the intermediate translation model as the context and use the pseudo-parallel corpus constructed from monolingual data as the supplement of the training set. The BLEU5-SBP scoring results on the released test set are shown in Table 5. Among all constrained systems, the primary system (uc-2021-istic-primary-a) ranked fourth. Among all of the participated systems, uc-2021-istic-contrast-b ranked sixth, uc-2021-istic-contrast-c ranked seventh, and uc-2021-istic-primary-a ranked eighth.

Table 5. BLEU5-SBP scoring of Uyghur-to-Chinese track on released test set.

System	BLEU5-SBP
uc-2021-istic-primary-a (decoder combination model)	0.3495
uc-2021-istic-contrast-b (baseline + model ensembling)	0.352
uc-2021-istic-contrast-c (encoder combination + model ensembling)	0.35

In the Russian-to-Chinese translation evaluation task, the primary system (rc-2021-istic-primary-a) takes the target language sentences as the context, trains 30 epochs with the Transformer baseline system, and uses the last epoch checkpoint to decode. The contrast system 1 (rc-2021-istic-contrast-b) uses external 123,605 Russian-Chinese dictionary data and 55,504 bilingual parallel corpus data and the training process is the same as the primary system. The BLEU5-SBP scoring results on the released test set are shown in Table 6. Among all of the participated systems, rc-2021-istic-contrast-b ranked second, and rc-2021-istic-primary-a ranked third.

Table 6. BLEU5-SBP scoring of Russian-to-Chinese track on released test set.

System	BLEU5-SBP
rc-2021-istic-primary-a (baseline)	0.069
rc-2021-istic-contrast-b (baseline + dictionary)	0.1077

The results show that: (1) Model averaging, model ensembling, and multi-encoder system combination are helpful to improve translation quality; (2) The construction of pseudo parallel corpus by monolingual data back translation is conducive to the improvement of translation quality; (3) The accuracy of data pre-processing has a great influence on the quality of translation; (4) The method of multi-dimensional and multi similarity fusion is helpful to filter the corpus and select higher quality parallel sentence pairs.

5 Conclusions

This paper introduces the main technologies and methods of ISTIC in CCMT '2021. To sum up, our model is constructed on the Transformer architecture of self-attention mechanism and context-based system combination method. In the aspect of data pre-processing, we explore several corpus filtering methods. In the process of translation output, the strategies of the model averaging, model ensemble are adopted. In the process of corpus filtering, Elasticsearch is used for similar corpus filtering. Experimental results show that these methods can effectively improve the quality of translation. For machine translation tasks of low resource language, adding external dictionaries and parallel corpus can effectively improve the translation performance.

Due to the limited time, many methods have not been tried in this evaluation. Some problems have been found in the evaluation process, and the translation model still has a lot of room for improvement. In the future, we hope to learn more advanced technology and contribute to the research of machine translation.

References

1. Vaswani, A., et al.: Attention is all you need. In: Advances in Neural Information Processing Systems 30 (NIPS 2017), pp. 5998–6008 (2017)
2. Zhang, J., Zong, C.: Neural machine translation: challenges, progress and future. Sci. China Technol. Sci. **63**(10), 2028–2050 (2020). https://doi.org/10.1007/s11431-020-1632-x
3. Voita, E., et al.: Context-aware neural machine translation learns anaphora resolution. In: Proceedings of the 56th Annual Meeting of the Association for Computational Linguistics, pp. 1264–1274 (2018)

4. Li, B., et al.: Does multi-encoder help? a case study on context-aware neural machine translation. In: Proceedings of the 58th Annual Meeting of the Association for Computational Linguistics, pp. 3512–3518 (2020)
5. Elasticsearch. https://github.com/elastic/elasticsearch. Accessed 25 May 2021
6. Claeskens, G.: Model Selection and Model Averaging. Cambridge University Press, Cambridge (2008). https://doi.org/10.1017/CBO9780511790485
7. Lutellier, T., et al.: CoCoNuT: combining context-aware neural translation models using ensemble for program repair. In: Proceedings of the 29th ACM SIGSOFT International Symposium on Software Testing and Analysis, pp. 101–114. Association for Computing Machinery, New York (2020)
8. Ott, M., et al.: fairseq: a fast, extensible toolkit for sequence modeling. In: Proceedings of the 2019 Conference of the North American Chapter of the Association for Computational Linguistics (Demonstrations), pp. 48–53 (2019)
9. Fairseq. https://github.com/pytorch/fairseq. Accessed 15 May 2021
10. Provilkov, I., et al.: BPE-dropout: simple and effective subword regularization. In: Proceedings of the 58th Annual Meeting of the Association for Computational Linguistics, pp. 1882–1892 (2020)
11. Sennrich, R., et al.: Neural machine translation of rare words with subword units. In: Proceedings of the 54th Annual Meeting of the Association for Computational Linguistics (Volume 1: Long Papers), pp. 1715–1725 (2016)
12. Wei, J., Liu, W., Wu, Z., Pan, Y., He, Y.: ISTIC's neural machine translation system for IWSLT 2020. In: Proceedings of the 17th International Conference on Spoken Language Translation, pp. 158–165. Association for Computational Linguistics (2020)
13. Urheen. https://www.nlpr.ia.ac.cn/cip/software.html. Accessed 15 May 2021
14. Polyglot. https://github.com/aboSamoor/polyglot. Accessed 15 May 2021
15. Papineni, K., et al.: BLEU: a method for automatic evaluation of machine translation. In: Proceedings of the 40th Annual Meeting of the Association for Computational Linguistics, pp. 311–318 (2002)

BJTU's Submission to CCMT 2021 Translation Evaluation Task

Xiang Li, Xuanxuan Wu, Shuaibo Wang, Xiaoke Liang, Li Yin, Hui Huang, Yufeng Chen, and Jin'an Xu[✉]

School of Computer and Information Technology, Beijing Jiaotong University, Beijing, China
{xiang.li,19120414,20120419,20125185,20125265,18112023,chenyf, jaxu}@bjtu.edu.cn

Abstract. This paper presents the systems developed by Beijing Jiaotong University for the CCMT2021 evaluation tasks. We joined four translation tasks of Chinese-English, English-Chinese, Uyghur-Chinese, Tibetan-Chinese. In all directions, we build our system based on transformer architecture and Dynamic-Conv. Additionally, we apply Byte Pair Encoding (BPE) to all translation tasks to resolve the out-of-vocabulary (OOV) problem. We also adopt some techniques that have been proven effective recently in academia, such as data augmentation, finetuning, model ensemble and reranking. Experiments show that our machine translation systems achieved high accuracy on all directions.

Keywords: Neural machine translation · Data augmentation · Finetuning · Model ensemble · Reranking

1 Introduction

This paper introduces in detail the submission of Beijing Jiaotong University to the translation evaluation task in the 17-th China Conference on Machine Translation (CCMT2021). We participated in both directions of Chinese-English translation tasks from the news field and two minority language translation tasks Tibetan-Chinese translation from government literature and Uyghur-Chinese translation from the news field.

In these directions, we built our system based on five different architectures, the first one is solely based on attention mechanisms, namely the Transformer-base model [11]. We broadened Transformer with bigger hidden dimensions and more attention headers to better extract features from source segments, which is named as Transformer-big. We also tried to augmented the encoder layers to extract more semantic information from the source which is named as Transformer-deep. Transformer-big and Transformer-deep are proved to outperform Transformer-base model in most cases [12]. Additionally, we also tried to substitute the self-attention layer with lightweight convolution, providing us with another different model to use when doing model ensemble [14].

© Springer Nature Singapore Pte Ltd. 2021
J. Su and R. Sennrich (Eds.): CCMT 2021, CCIS 1464, pp. 117–124, 2021.
https://doi.org/10.1007/978-981-16-7512-6_10

Additionally, we applied sub-word segmentation to both languages to resolve the out-of-vocabulary problem [9]. To deal with the scarcity of training data, we created massive synthetic data using back translation based on monolingual Chinese data [8]. To make full use of the translation knowledge learned by other decoding models, knowledge distillation is used to integrate various knowledge into one model [3].

The in-domain finetuning is very effective in our experiments and especially, we used a boosted finetuning method for Chinese→English and English→Chinese tasks. We also take advantage of the combination methods to further improve the translation quality.

We also applied two model ensemble techniques, namely model averaging and model ensemble, to leverage multiple models to further improve the result [1,7]. To alleviate unbalanced output and error accumulation during left to right decoding, we performed reranking on the top-k outputs based on z-Mert algorithm [4].

2   Data

2.1   Chinese-English

We use all available data provided by CCMT'21 and WMT'21, which contain 28.6M bilingual sentence pairs and 100M Chinese Monolingual data and 120M English Monolingual data. We apply the following procedures to preprocess the data:

1. Remove illegal UTF-8 characters and replace control characters with a single space.
2. Convert Traditional Chinese sentences into Simplified Chinese.
3. Apply Unicode NFKC normalization.
4. Remove duplicated sentence pairs.
5. Keep parallel sentences with a length ratio between 0.7-2.2.
6. Truecase[1] the English corpus.

For the new corpus "ParaCrawl v7.1" in WMT'21, there are plenty of noisy sentence pairs. We have trained a baseline model with Transformer-base to filter out the noisy pairs with SacreBLEU lower than 35.0.

2.2   Uyghur→Chinese

We use the parallel data provided by CCMT'21, which contains 0.17M pairs. We cleaned the provided training data accords to two criteria, namely the length ratio of source to target for each sentence pair, and the average length of source sentence and target sentence.

---

[1] https://github.com/moses-smt/mosesdecoder.

2.3 Tibetan→Chinese

We use all available data provided by CCMT'21, which has 0.15M parallel sentences. However we have not used the devset provided by CCMT'21, we randomly sample 1k sentences in parallel sentences as our devset and the rest of the available data is used as training data. We apply Unicode NFKC to normalize the data. For Tibetan word segmentation, we build a vocabulary which consists of 140k words, and use Bidirectional-Maximum Matching algorithm.

3 Model

As we explained before, we combined four different architectures in our work, namely Transformer-base, Transformer-deep, Transformer-big and Light-Conv.

Transformer-Base. Transformer is a completely attention-based structure for dealing with problems related to sequence models [10], such as machine translation. The Transformer model does not use any CNN or RNN structure, capable of working in the process of highly parallelization, so the training speed is very fast while improving the translation performance. Transformer-base is the naive version of transformer.

Transformer-Deep. The performance of Transformer can be improved by increasing the number of layers in the encoder. We follow to use deep Transformer. To address the vanishing-gradient problem in deep Transformer, we use the post-layer normalization instead of the pre-layer normalization. In Chinese-English directions, we adopt this model which has great performance.

Transformer-Big. In some cases, Transformer-deep does not perform better than big, which have a fewer parameters than the former. Therefore, for the stage of training and inference, Transformer is faster than Transformer-deep.

LightConv. Lightweight convolution uses the prototype of deep (separable) convolution in CV domain, which greatly reduces the number of parameters and reduces the complexity by sharing parameters in the channel dimension. On the basis of light weight, dynamic convolution is proposed, where the weight of CNN is calculated dynamically from the input feature. The Dynamic-Conv model is proved to be competitive with Transformer model in many scenarios.

4 Method

4.1 Data Augmentation

Back-Translation. We augment the training data by exploring the monolingual corpus using back translation. Specifically, we select target monolingual corpus which has the same size as the training corpus and then translate them back into the source language using target-to-source (T2S) models. We merge the synthetic data with the bilingual data to train our models. We also add noise to the translated sentences to further improve the performance namely Noisy Back-Translation.

Knowledge Distillation. The existing translation model decodes from left to right (L2R), and from source to target (S2T). In order to make full use of the translation knowledge learned by other decoding models, knowledge distillation is adopted to improve the translation performance. Knowledge distillation is a method for knowledge transfer, where the prediction distribution of teacher model is used to guide the parameter learning of student model. In our submission, the following three teacher models are trained first:

1. The translation model decodes from source to target and from left to right (L2R).
2. The translation model decodes from source to target and from right to left (R2L).
3. The translation model decodes from target to source and from left to right (T2S).

After obtaining the above three translation models, we use the method of sentence level knowledge distillation to decode the training data and get their respective decoding results, and form the bilingual sentence pairs of knowledge distillation with their respective input sentences. In this evaluation, we mixed the knowledge distillation bilingual sentence pairs with the original training data. In this way, in mixed bilingual data, in addition to the original training data, it also contains the prediction results of the respective teacher models. Finally, the student model is retrained with mixed training data.

4.2   Model Average

Because of the mismatch of BLEU and MLE Loss in the final convergence stage, we have applied the Model Average method to average the parameters from the last several checkpoints. We have found that Model Average works on Uyghur→Chinese and Tibetan→Chinese but makes no sense in Chinese-English.

4.3   Finetune

Finetuning [2] with in-domain data can bring huge improvements. We also use development set as the in-domain dataset. The source side of newsdev2017, newstest2017 and newstest2018 are composed of two parts: documents created originally in Chinese and documents created originally in English. We split these datasets into original Chinese part and original English part according to tag attributes of SGM files. For Chinese-English translation, we use CWMT2008,

CWMT2009 and original Chinese part of newsdev2017, newstest2017, newstest2018 and newstest2020 as the in-domain dataset. For English-Chinese translation, we use original English part of newsdev2017, newstest2017, newstest2018 and newstest2020 as the in-domain dataset. During finetuning, we use a larger dropout rate, a smaller constant learning rate and batch size. The parameters are updated after each epoch, which is enabled by using gradient accumulation.

4.4   Model Ensemble

Ensemble is a well-known technique to combine different models for stronger performance. We utilize the frequently used method for ensemble, which calculates the word level averaged log-probability among different models during decoding. On account of the model diversity among single models has a strong impact on the performance of ensembling models, we combine single models that have different model architectures (Transformer-base, Transformer-big, Transformer-deep, Transfomer-deepbig, Light-conv, Dynamic-conv). We also try to use Transductive Ensemble Learning (TEL) [13] to replace ensemble. TEL is a technique utilizing the synthetic test data (consists of original source sentences and translations of target-side) of different models to finetune a single model.

4.5   Reranking

Neural machine translation models are usually decoded from left to right, and are faced with the problem of unbalanced output and error accumulation. In the process of translation generation, if there are errors in the first few moments, it is difficult to produce correct results in the following. To some extent, this problem can be alleviated by increasing the space of beam search. However, since we only select the sentence with the highest prediction probability as the final output, the increase of searching space will not bring significant benefits, and even bring some performance losses. Therefore, this paper uses the method of reranking. In this paper, several feature models are trained to grade the candidate translation. The feature models include the R2L model, L2R model, T2S model and language model scores. Word-penalty is also included to penalize too short output, which is the length of each candidate. After that, z-Mert [15] is used to rerank the candidate translations, and the translation with the highest score is selected as the final output translation.

5   Experiment

5.1   Chinese→English

We use the PyTorch implementation of open-source toolkit fairseq [5] to conduct all experiments. To enable open vocabulary, we learn 32K BPE operations separately on Chinese and English texts using subword-nmt toolkit. We set Chinese vocabulary size of 40k and English vocabulary size of 32k. All models are trained on Tesla-V100. Table 1 shows the results of Chinese-English Translation on newstest2019 dataset. All methods we used can bring substantial improvement over the baseline system. Applying data augmentation methods improve the baseline system by 2.3 BLEU score. Finetuning is the most effective approach. With transductive ensemble on newstest2019 our model has achieved 40.12.

Table 1. BLEU evaluation results on the newstest2019 Chinese-English test set

Settings	Transformer-big	Transformer-deep	Lightconv
Baseline	27.72	28.14	27.11
+ data augment	30.12	30.07	29.88
+ finetuning	39.32	38.55	38.32
Ensemble	40.12		

5.2 English→Chinese

We have the same preprocessing setting with the Chinese→English direction. And all the models are trained on RTX 1080Ti. However, the back-translation does not work, therefore we just apply noisy back-translation. Our results are depicted as Table 2 where finetuning in English→Chinese does not have the same improvement.

Table 2. BLEU evaluation results on newstest2019 English-Chinese test set

Settings	Transformer-big	Transformer-deep	Lightconv
Baseline	36.96	35.75	–
+ data augment	37.34	36.66	37.43
+ finetuning	38.47	37.46	38.77
Ensemble	39.81		

5.3 Uyghur→Chinese

In Uyghur, we adopt fast-align and kenLM to select the monolingual data. We then back translate the monolingual sentences to generate the twice size of the parallel data. Finally we combine the parallel data and the pseudo-parallel data.

In our experiment, we utilize the BPE-Dropout [6] as a method of data augmentation with the dropout rate 0.1. BPE-Dropout performs well on the mini-scale dataset.

Table 3. BLEU evaluation results on CCMT'21 Uyghur-Chinese dev set

Settings	Transformer-base	Transformer-big	Transformer-deepbig	Dynamic-conv
Baseline	41.12	40.09	41.03	43.91
+ data augment	43.96	43.85	43.75	45.17
+ finetuning	44.65	45.03	45.79	45.22
Ensemble	48.02			

Table 4. Result of BPE-dropout in Uyghur-Chinese.

Models	BPE	BPE-dropout
Transformer-base	41.12	43.96
Dynamic-conv	43.91	45.17

Table 5. BLEU evaluation results on CCMT'21 Tibetan-Chinese dev set.

Settings	Transformer-base	Transformer-big	Dynamic-conv
Baseline	46.83	46.48	46.76
+ data augment	47.34	46.91	46.97
+ finetuning	48.50	47.62	48.90
Ensemble	51.09		
Rerank	54.35		

5.4   Tibetan→Chinese

Table 5 shows the result of Tibetan→Chinese that reranking has improved 3.3 BLEU score, which does not make sense in Chiense→Englsih, English→Chinese and Uyghur→Chinese.

6   Conclusion

In this paper, we described our submission in four translation evaluation projects including Chinese to English, English to Chinese, Tibetan to Chinese and Uyghur to Chinese. In all directions, we build our system based on six different architectures, namely Transformer-base, Transformer-big, Transformer-deep, Transformer-deepbig and Dynamic-Conv. Finally, we obtain substantial improvements combining these methods. Our training strategies including back-translation, knowledge distillation, model ensemble and reranking have good performance in these tasks.

Acknowledgement. The research work descried in this paper has been supported by the National Key R&D Program of China 2020AAA0108001 and the National Nature Science Foundation of China (No. 61976015, 61976016, 61876198 and 61370130). The authors would like to thank the anonymous reviewers for their valuable comments and suggestions to improve this paper.

References

1. Chen, H., Lundberg, S., Lee, S.I.: Checkpoint ensembles: Ensemble methods from a single training process. arXiv preprint arXiv:1710.03282 (2017)
2. Chu, C., Dabre, R., Kurohashi, S.: An empirical comparison of domain adaptation methods for neural machine translation. In: ACL, pp. 385–391 (2017)
3. Kim, Y., Rush, A.M.: Sequence-level knowledge distillation. In: EMNLP, pp. 1317–1327 (2016)
4. Olteanu, M., Suriyentrakorn, P., Moldovan, D.: Language models and reranking for machine translation. In: Proceedings on the Workshop on Statistical Machine Translation, pp. 150–153 (2006)
5. Ott, M., et al.: fairseq: a fast, extensible toolkit for sequence modeling. In: NAACL, pp. 48–53 (2019)
6. Provilkov, I., Emelianenko, D., Voita, E.: Bpe-dropout: Simple and effective subword regularization. arXiv preprint arXiv:1910.13267 (2019)
7. Rokach, L.: Ensemble-based classifiers. Artif. Intell. Rev. **33**(1–2), 1–39 (2010). https://doi.org/10.1007/s10462-009-9124-7
8. Sennrich, R., Haddow, B., Birch, A.: Improving neural machine translation models with monolingual data. In: ACL, pp. 86–96 (2016)
9. Sennrich, R., Haddow, B., Birch, A.: Neural machine translation of rare words with subword units. In: ACL, pp. 1715–1725 (2016)
10. Sutskever, I., Vinyals, O., Le, Q.V.: Sequence to sequence learning with neural networks. In: NIPS, pp. 3104–3112 (2014)
11. Vaswani, A., Shazeer, N., et al.: Attention is all you need. In: NIPS, pp. 5998–6008 (2017)
12. Wang, Q., et al.: Learning deep transformer models for machine translation. arXiv preprint arXiv:1906.01787 (2019)
13. Wang, Y., Wu, L., Xia, Y., Qin, T., Zhai, C., Liu, T.Y.: Transductive ensemble learning for neural machine translation. In: AAAI, pp. 6291–6298 (2020)
14. Wu, F., Fan, A., Baevski, A., Dauphin, Y.N., Auli, M.: Pay less attention with lightweight and dynamic convolutions. arXiv preprint arXiv:1901.10430 (2019)
15. Zaidan, O.: Z-MERT: a fully configurable open source tool for minimum error rate training of machine translation systems. Prague Bull. Math. Linguist. **91**, 79–88 (2009)

Author Index

Printed in the United States
by Baker & Taylor Publisher Services